# The Sassafras Science Adventures

# Volume 4: Earth Science

Johnny Congo & Paige Hudson

# THE SASSAFRAS SCIENCE ADVENTURES
## VOLUME 4: EARTH SCIENCE

First Printing 2016
Copyright @ Elemental Science, Inc.
Email: info@elementalscience.com

ISBN: 978-1-935614-43-2
Cover Design by Paige Hudson & Eunike Nugroho
Illustrations by Eunike Nugroho (be.net/inikeke)

Printed In USA For World Wide Distribution

No part of this book may be reproduced or transmitted in any form or by any means, electronic or mechanical, including photocopying, recording, or by means of any information storage and retrieval system, without permission in writing from the authors. The only exception is brief quotations in printed reviews.

**For more copies write to :**
Elemental Science
610 N Main St #207
Blacksburg, VA 24060
info@elementalscience.com

## Dedication

We dedicate this book to our parents – those who are with us and those who are not. Your love and support has helped us turn our dream of writing something like the Sassafras Science series into a reality.

# Table of Contents

The Sassafras Guide to the Characters ix

Chapter 1: Embarking on Earth Science 1

   *Entering the Territory of the Guardian Beast*    *1*
   *A Photosyntastic Program*    *8*

Chapter 2: O-o-o-o-klahoma 17

   *Where the wind comes sweeping down the plains*    *17*
   *Easterlies, Westerlies, and Forgotten Memories*    *25*

Chapter 3: Lucille's First Rodeo 35

   *Damaging Downbursts*    *35*
   *Thunderstone's Tornado*    *45*

Chapter 4: The Congolese Jungle Treasure Hunt 55

   *Carver's Rain Takes the Lead*    *55*
   *Moving Monsoons*    *64*

Chapter 5: The Search for the Giant Bonobo Diamond 73

   *Thundering Pygmy Warriors*    *73*
   *Flooded Findings*    *80*

Chapter 6: Parachuting into Patagonia 93

   *Snowy Set Downs*    *93*
   *Icy Impositions*    *102*

The Sassafras Science Adventures

## Chapter 7: Out of the Office — 113

*Frost Quake!* — 113
*Seasonal Shifts* — 123

## Chapter 8: The Gobi Desert — 131

*Is it day or night?* — 131
*Swirling Sandstorms* — 139

## Chapter 9: Avargatom Challenges — 147

*Desperate Drought* — 147
*Oasis in the Trials* — 156

## Chapter 10: Wolves in Pakistan — 163

*Ascending through the Atmosphere* — 163
*Capturing Clouds* — 172

## Chapter 11: The Lost One is Found — 181

*Out of the Alto Clouds* — 181
*Into the Stratus Market* — 189

## Chapter 12: Back in Alaska — 201

*A Break in the Water Cycle* — 201
*PB and J with a Side of Fog* — 210

## Chapter 13: The Forget-O-Nator — 221

*Competing Cycles* — 221
*Phosphorus Predicaments* — 230

## Chapter 14: A Watery Landing... Again — 241

*Careening towards Coral* — 241
*Drifting in Currents of Confusion* — 250

The Sassafras Science Adventures

## Chapter 15: The Threat of Thaddaeus    259
- **Oceanic Occurrences**    *259*
- **Typhoon Thaddaeus**    *268*

## Chapter 16: Quick! To Switzerland!    279
- **Grim Groundwater**    *279*
- **I Spy: Waterfall**    *289*

## Chapter 17: Tracking down Bogdanovich    299
- **River Reconnaissance**    *299*
- **Lakeside Lasso**    *308*

## Chapter 18: The End of Earth Science    321
- **Bonus Data**    *321*
- **Jumbled Geology**    *330*

The Sassafras Science Adventures

# MAKE THE MOST OF YOUR JOURNEY WITH THE SASSAFRAS TWINS!

Add our activity guide, logbook, or lapbooking guide to create a full science curriculum for your students!

*The Sassafras Guide to Earth Science* includes chapter summaries and an array of options that coordinate with the individual chapters of this novel. This guide provides ideas for experiments, notebooking, vocabulary, memory work, and additional activities to enhance what your students are learning about our Earth!

*The Official Sassafras SCIDAT Logbook: Earth Science Edition* partners with the activity guide to help your student document their journey throughout this novel. The logbook includes their own SCIDAT log pages as well as climate sheets and an earth science glossary.

*Lapbooking through Earth Science with the Sassafras Twins* provides a gentle option for enhancing what your students are learning about our Earth through this novel. The guide contains a reading plan, templates, and pictures to create a beautiful lapbook on earth science, vocabulary, and coordinated scientific demonstrations!

## VISIT SASSAFRASSCIENCE.COM TO LEARN MORE!

THE SASSAFRAS SCIENCE ADVENTURES

# The Sassafras Guide to the Characters

## Throughout the Book*

★ **Blaine Sassafras** – The male Sassafras twin, also known as Train. So far this summer, he has swung upside down in the trees, fallen out of a heliquickter, and lost his phone multiple times.

★ **Tracey Sassafras** – The female Sassafras twin, also known as Blaisey. So far this summer, she has been kidnapped by an Amazonian tribal leader, caught in a rockslide, and trapped inside a box.

★ **Cecil Sassafras** – The Sassafras twins' crazy, but talented uncle. He is eccentric and messy, but his brilliant mind co-invented the invisible zip-lines and several other contraptions.

★ **President Lincoln** – Uncle Cecil's lab assistant, who also happens to be a prairie dog. He doesn't say much, but his talent has been used to create amazing presentations, fix glitches, and co-invent the invisible zip-lines.

★ **The Man with No Eyebrows** – He has no eyebrows and seems to be trying to sabotage the twins at every stop. He has broken into Cecil's lab and has been spying on Cecil's every move.

(*Note – These characters also appeared in the first three volumes of *The Sassafras Science Adventures* series.)

## Cecil's Neighborhood (Chapters 1 & 18)

★ **Old Man Grusher's Dog** – Also known as the "guardian beast". This miniature poodle loves to chase Cecil Sassafras.

★ **Mrs. Pascapali (paz-kah-pah-LEE)** – She is Uncle Cecil's neighbor who lives at 1106 North Pecan Street.

★ **Preston** – He is the squeaky and skinny teenaged clerk of the Left-handed Turtle Market.

## Oklahoma City (Chapters 2 & 3)

★ **Sylvia Thunderstone** – The twins' local expert for their time in the Oklahoma prairie. She is a native Oklahoman and the meteorologist in charge of Lucille, the storm-chasing vehicle.

- **Sylvester Hibbel (Doc)** – He is a traveling salesman, cowboy enthusiast, and inventor of several medicinal elixirs. The twins first met him during their anatomy leg.
- **Jayman** – He is a friend and colleague of Dr. Thunderstone. He is responsible for relaying weather information from the main station to Lucille.

## THE CONGO (CHAPTERS 4 & 5)

- **Carver Brighton** – The twins' local expert for their time in the Democratic Republic of the Congo. He is a professor of geochemistry who is serving as the scientific expert for the Giant Bonobo Diamond Treasure Hunt.
- **Garfield T. Wellington the Fourth** – He is the benefactor of the Giant Bonobo Diamond Treasure Hunt.
- **Stuart Dimsley** – He is a long-time rival of Carver Brighton's and a professor of cultural studies. He serves as the cultural expert for the Giant Bonobo Diamond Treasure Hunt.
- **Bakaza (bah-KAH-zah)** – He is the Congolese guide and trailblazer for the Giant Bonobo Diamond Treasure Hunt.
- **Chief Wazabanga (wah-zah-BANG-ah)** – He is the chief of the pygmy warriors that the treasure hunters run into while on their search.

## PATAGONIA (CHAPTERS 6 & 7)

- **Hawk Talons** – He is the twins' local expert for their Patagonia leg and host of the "Out of the Office" TV show. He is an adventurer, scientist, survivalist, and member of Antarctica's Special Forces.
- **Ted** – He is one of the workers at the Q.B. Cubicles office and is very reluctantly participating in the latest episode of "Out of the Office."
- **Mitchell** – He is also an employee at Q.B. Cubicles and is participating in the latest episode of "Out of the Office." He is excited to be a part of the challenge, but is not the sharpest tool in the shed.
- **Tammy** – She is another one of the workers from Q.B. Cubicles

participating in the latest episode of "Out of the Office".
* **Barbara** – She is a bit of a hypochondriac who is also from Q.B. Cubicles and participating in the latest episode of "Out of the Office."

## Mongolia (Chapters 8 & 9)

* **Ganzorig (gan-ZOR-ig) Buri** – He is the twins' local expert for their time in the Mongolian Desert. Ganzorig is a college student returning home for his summer break.
* **Solongo (so-LONG-o)** – A close friend of Ganzorig's. She is from the same Buri village as he is.
* **Avargatom (AH-var-gat-um)** – A band of large-statured raiders who travel throughout the Mongolian desert stealing from small villages.
* **Dariin (dar-EEN)** – One of Ganzorig's brothers.
* **Khulan (KOO-lan)** – One of Ganzorig's brothers.

## Pakistan (Chapter 10 & 11)

* **Atif (A-teef) Jilani (JEE-lahn-nee)** – He is the quiet, steady, and knowledgeable local expert for the twins' time in Pakistan. He is known to everyone as the Shepherd.
* **Javeria (ha-ver-EE-ah)** – She is an orphan and one of the Shepherd's apprentices.
* **Aazmi (ahz-MEE)** – He is an orphan and one of the Shepherd's apprentices.
* **Tariq (Tah-REEK)** – He is an orphan and one of the Shepherd's apprentices.
* **Qaiser (KA-zer) Qazi (KA-zee)** – An infamous thief known as the Raider.
* **Naveed (nah-VEED)** – A former student of the Shepherd's.
* **The Magistrate** – A governor in the Karakoroam region of Pakistan; his word is law.

## Alaska (Chapter 12 & 13)

* **Summer Beach** – The loveable, excitable scientist and Alaskan

local expert. Blaine and Tracey have gotten to know and love her throughout their adventure.
* **Ulysses S. Grant** – Summer's lab assistant, who happens to be an arctic ground squirrel. He is the inventor of the robot squirrels.
* **Yotimo** – The stoic Alaskan native who saved Tracey from a polar bear during the twins' zoology leg.
* **Skeeter and Tina Romig (ROOM-ig)** – Friends of Summer's who are science teachers that recently moved to Alaska.

## THE PACIFIC OCEAN (CHAPTERS 14 & 15)

* **Billfrey Battaballabingo** – The twins' local expert for their time in the Pacific Ocean. He is a marine biologist who has spent far too much time alone on the Western Garbage Patch.
* **Cantankerous Carl** – Billfrey's ladder-man companion.
* **Sticky Fingers Stevie** – Billfrey's coat-rack-man companion.
* **Mr and Mrs. Osodarling** – The broom and mop couple who keep Billfrey company.
* **Ig** – Billfrey's mannequin companion.
* **Peach Beard** – The not-too-bright captain of the P.R.O. Pirates whom the twins first met on their zoology leg.

## SWITZERLAND (CHAPTER 16 & 17)

* **Evan DeBlose** – The Triple S agent who serves as the twins' local expert for their time in Switzerland, a.k.a. Agent Pork.
* **Jorgen Wuthrich** – Triple S agent and DeBlose's partner, a.k.a. Agent Beans.
* **Yuroslav Bogdanovich** – He is the rogue scientist and evil villain who travels around Europe to carry out his evil schemes. The twins first met him during their botany leg.
* **Adriana Archer** – Triple S agent who is goes by the name Agent Mac. She is partners with Agent Zwyssig.
* **Gottfried Zwyssig (zzz-WHY-zig)** – Triple S agent who goes by the name Agent Cheese. He is partners with Agent Archer.
* **Captain Marolf** – Head of the Triple S Agency.
* **Q-Tip** – The Triple S's expert in technologizing.

# Volume 4

# Earth Science

# Volume 4: Earth Science

## Chapter 1: Embarking on Earth Science
### *Entering the Territory of the Guardian Beast*

Somehow, they had made it without being detected.

They had taken quiet steps. They had uttered no words.

They had executed the journey with utmost caution. Their mission had been a success.

They had made it to their destination without being seen, heard, or smelled by the guardian beast, despite the scent of their bacon-laden breakfast still lingering on their clothing.

The three were all now standing in front of the target location: The Left Handed Turtle Neighborhood Market.

Cecil, Blaine, and Tracey Sassafras stood victoriously in front of the automated glass doors. Cecil was quite a bit more elated than the twins were, probably because he had been terrified about making the journey. Blaine and Tracey, on the other hand, had not been scared at all. They didn't see what the big deal was about walking the few blocks from their uncle's house to the neighborhood supermarket. Yes, on the way you had to walk past Old Man Grusher's house. And on the porch of that house there was usually a dog, but the dog was just a miniature poodle for goodness sake. It wasn't the "guardian beast" that Uncle Cecil liked to call it.

Blaine and Tracey were visiting their uncle for the summer due to their failing grades in science. They had gotten fantastic grades in every other subject, but not in science, which they despised in school. When their parents had found out about Blaine and Tracey's 'F's', they formulated a plan to send their twelve-year-olds away for the summer to work on "their science problem," as they had said. Uncle Cecil, who was their dad's brother, happened to be a pseudo-famous research scientist. So Blaine and Tracey's parents figured if anybody could help their children turn their science grades around, it would be Cecil. And, so far, their parents' plan was working splendidly.

Over the course of the last few weeks or so that they had spent with their uncle, Blaine and Tracey had gone from despising science to actually loving it! As the weeks went by, the twins were beginning to consider themselves as defenders of science. Uncle Cecil was absent-minded and more than a little wacky, but Blaine and Tracey had truly grown to love and appreciate him along the way.

Their uncle's way of enabling them to learn about science was absolutely out of this world! Cecil, along with a prairie dog lab assistant named President Lincoln, had invented invisible zip-lines. And not only were the lines invisible, but they were global and fast—real fast. These unbelievably amazing lines could zip the twins to any location on the planet at the speed of light!

All they needed to ride the zip-lines was a harness and a specially designed three-ringed carabiner. One ring on the carabiner was for longitude coordinates, one ring was for latitude coordinates, and the last ring was to secure the harness to the invisible lines. When they first heard about the mode of travel it sounded too good to be true, but it wasn't—Blaine and Tracey had been experiencing this phenomenon for weeks now.

At each location they zipped to, the twins met a local expert who would help them study several scientific topics. It was this hands-on, face-to-face, experiential way of learning that had won over the Sassafras twins' hearts.

So far, they had zipped through the subjects of zoology, anatomy, and botany. Now, this very morning, they were anticipating the start of another zip-riffic science subject. But first, Uncle Cecil said that they needed to pick up some groceries. So here they were now, standing in front of the Left-Handed Turtle.

The three Sassafrases stepped through the market's automatic front doors and were immediately greeted by a squeaky, cracking, teen-age voice, "Welcome to Left-Handed Turtle! Welcome to the Left-Handed Turtle! Welcome to the Left-Handed Turtle!"

"Good morning, Preston!" Cecil joyfully shot back to the young store clerk who had squeaked out salutations to the three on behalf of the store. The eldest Sassafras grabbed a shopping cart and skipped down the first aisle to commence shopping. The twins followed close behind.

"Why did he say, 'Welcome to the Left-Handed Turtle three times?'" Blaine wondered out loud.

"Who, Preston?" Cecil asked. "The front clerk is supposed to greet every individual that comes in with a warm Left-Handed salutation."

Both twins nodded and smiled at the same time. All the while they were thinking how this trip to the supermarket just highlighted the fact that so many of the quirks about Uncle Cecil's neighborhood were as delightfully strange as he was.

Cecil began whistling as skipped and zigzagged down the

market's aisles, haphazardly grabbing items and throwing them into his cart. The twins had a hard time keeping up with him as he speedily wound around the supermarket. Eventually, all three Sassafrases ended up back at the check-out stand with a huge, overflowing mound of mismatched groceries.

Preston, the squeaky and skinny teenaged clerk, painstakingly rang up all of Cecil's items and then sacked them in a big pile of blue plastic bags. "That will be three hundred seventy-six dollars and forty-two cents, Mr. Sassafras," the young man informed, his voice cracking twice during the sentence.

Cecil, who was wearing his usual white science lab coat, reached into the coat's side pocket for some cash with which to pay, but his hand came out empty. A look of slight panic formed on the scientist's face as he now began searching all the other pockets on his person. The pockets on his coat, his pants, and his shirt all came up empty.

"Did you forget your money again, Mr. Sassafras?" Preston asked, evidently not surprised at Cecil's current lack of funds.

"Oh, slippety slappety geeze little weeze. I sure did, Preston," Cecil sighed. "I left my bills and cents in my left shoe."

"Your left shoe?" Preston questioned. "Why don't you just take your shoe off now and pay for your groceries, then, sir?"

"Because I'm not wearing my left shoe."

"You're not wearing your left shoe?"

"No."

"Are you wearing your right shoe?"

"No, Preston, I'm wearing neither my right shoe nor my left shoe. I am wearing my house slippers that look like fuzzy bunnies because they are as soft and comfortable as cotton candy."

"Then where is your left shoe, sir?"

"My left shoe is in a fishbowl."

THE SASSAFRAS SCIENCE ADVENTURES

"In a fishbowl, sir?"

"Yes, a fishbowl. My left shoe is in the fishbowl, which is in the banana box in the parachute bag inside the zebra-striped dresser, which is in the mop closet under the stairs next to the living room."

Preston's acne spotted face now held a blank stare.

"I figure that's the safest place to keep cash," Cecil said plainly. He then switched to an apologetic tone. "Preston, I am so sorry, but can I leave my groceries here, go home, get my left shoe, and then return to settle the debt?"

"Sure, Mr. Sassafras, no problem." Preston chuckled kindly.

"Thank you, thank you, thankity thank you!" Cecil responded, as he rushed out of the Left-Handed Turtle, accompanied by his niece and nephew. He began making a beeline down the neighborhood sidewalk back toward his house, which was three blocks away at 1104 N Pecan Street. But after only a couple of dozen strides, he stopped abruptly and became as frozen as an ice sculpture.

"What's wrong, Uncle Cecil?" Tracey asked, concerned.

Cecil remained silent for a few more minutes and then finally managed to speak. "If we have to go home and get my left shoe, we will have to come all the way back here to the Left-Handed Turtle, which means we will have to pass . . . the guardian beast . . . two . . . more . . . times."

"Uncle Cecil, why are you so afrai—" Blaine started to say but was interrupted by a sharp elbow in the ribs from his sister.

"Don't worry, Uncle Cecil," Tracey comforted. "We will get by Old Man Grusher's dog—I mean, the guardian beast—with no problems. Blaine and I will be walking by your side the whole time."

Cecil nodded. He summoned up the courage, and then

started again toward home. The first two blocks were easy enough, and the pace was rather quick, but the third block required a left turn onto Pecan Street. Cecil's heartbeat quickened as his pace slowed to a tiptoe. They passed 1112 on the left and 1111 on the right—no sight of Old Man Grusher's miniature poodle yet.

Silently, cautiously, and slowly, they now passed 1110. Still, no sound of barking reached their ears. They moved forward slowly as they entered into what Uncle Cecil called "the guardian beast's territory." Old Man Grusher's house stood directly to their right. They could see the plastic golden house numbers—one, one, zero, and seven—glimmering in the morning sunlight. They had a clear view of the old man's porch—the throne room of the beast—but the dog was not there! The miniature poodle must have been in the backyard digging holes or maybe inside watching TV with Old Man Grusher.

The three kept their eyes trained to the right and continued on safely to Cecil's house.

He snickered to himself as they passed. Blaine, Tracey, and Cecil had been so fixated on Old Man Grusher's porch that they hadn't seen him standing silently in his front yard at 1108. He had stared at the three as they passed. And if they had looked his way, they would have seen a man wearing a wide-brimmed sun hat, holding a pair of shears that he was planning to use vigorously to cut small branches off one of the three sassafras trees in his front yard.

The wide-brimmed hat was there to conceal the fact that he had no eyebrows. And had they looked, they would not have known that he wanted to do much more than cut off the small branches off the sassafras tree. His real desire was to vigorously cut away the hopes and dreams of the three people walking past the front of his house.

He had first met Blaine and Tracy in Kenya on a safari tour with Nicholas Mzuri. He had watched them learn about lions and cheetahs. He had acted like he was a part of their tour group, but then he stole the group's jeep and left them marooned overnight out among the wild African animals! Unfortunately, with the help of Mzuri, the twins had survived Kenya. Then, they continued to move forward through many more science learning locations.

But he had also kept moving forward. He had found a way to access the same invisible zip-lines the twins did. He had been spotted by them multiple times since that first location.

The twins had no idea why he was so bent on their failure. They didn't really even have any idea who he was. Despite that, he had not stopped them from learning science. In fact, it seemed he had only enhanced their summer of learning by providing a battle for them to fight.

He guessed that knowing there was a malicious eyebrow-less man trying to thwart their efforts only made Blaine and Tracey want to learn all the more. Again and again, he had failed to stop them. Now they felt victorious over him. Any fear they had felt was replaced with confidence in their role as defenders of science. He grimaced at the thought, but straightened up as a wicked sneer curled up his lips.

That confidence would definitely take a hit if the Sassafras twins knew of everything he was up to. There was so much they were still in the dark about. They didn't know that he lived only two doors down from their uncle. They didn't know that he had hidden tiny cameras all over Cecil's house, which enabled him to watch all their moves and hear all their conversations. They didn't know that he had a machine in his basement that could wipe away an individual's memory. They didn't know that he was planning to kidnap both of them, put them in this machine, and erase all the knowledge they had gained. They didn't know that he was motivated by revenge and that he despised their uncle more than anyone on earth.

He was bound and determined to destroy anything that Cecil Sassafras loved. And right now, at the top of that list, was this zip-lining, science-learning, summer project that Cecil had planned for his niece and nephew. Yes, the man the three had unknowingly passed at 1108 North Pecan Street was up to a lot more than the Sassafrases knew about. His heart was darker and more devious than they might dare to imagine. And in the end, he was sure that he would be victorious!

### *A Photosyntastic Program*

The three Sassafrases reached Cecil's front porch safely, and before they knew it, they were all careening down into darkness. For most people, careening down into darkness would not seem safe, but it was pretty normal for the Sassafrases. They knew they were on a slide that started at a trap door located on Uncle Cecil's front porch and ended in his basement, where something wacky and science-y was always happening.

The twins had entered the lab through the trap door slide before, and now it was just plain fun for them. At the end of the slide, the three landed with soft thuds on a pile of old pillows. Cecil quickly performed three somersaults and a crashing cartwheel, picked himself up, and bounded across the cluttered basement. The twins followed, thinking Cecil would take a turn for the stairs, where he would go up and get his left shoe. But instead, he ran straight for his computer desk, where a happy looking prairie dog

VOLUME 4: EARTH SCIENCE                                                    9

was standing there waiting.

The prairie dog was none other than President Lincoln, Uncle Cecil's lab assistant and inventor extraordinaire. Just like Blaine and Tracey were used to the trap door slide, they were also accustomed to the fact that Uncle Cecil's lab assistant was an animal. They still weren't sure, though, if President Lincoln was really capable of all the things Cecil gave him credit for. Nonetheless, President Lincoln was a pretty amazing animal.

Upon the humans' arrival, the animal tapped at the computer keyboard and, immediately, a picture of the prairie dog could be seen on a big screen on the wall behind the computer. Above the prairie dog, the screen said, "President Lincoln's ever-so-brief presentation about botany."

Uncle Cecil smiled a huge excited smile and clapped his hands. He loved these presentations that his lab assistant gave. Blaine and Tracey liked them as well. The presentations reviewed science they had worked so hard to learn, but the videos also

THE SASSAFRAS SCIENCE ADVENTURES

usually preceded the start of a fresh scientific subject and a brand-new adventure around the world!

Cecil, as always, served as the exuberant emcee of the presentation. "I am so excited to begin this photosyntastic review with you all!" he started, before he joyfully read the presentation's title. President Lincoln then brought up the next page.

"First up, we have non-flowering plants!" the messy, redheaded scientist exclaimed. "Non-flowering plants are plants that do not have flowers. These are simple plants, like mosses and ferns, which have spores, and conifers, which reproduce through naked seeds."

The picture that appeared on the screen with the text was one of a Sitka spruce. The twins had taken the image of the coniferous evergreen tree with their smartphones while listening to Park Ranger Brock Hoverbreck. He had been their local expert in the foggy forest of Northern California, where the twins had joined a crew hunting for Bigfoot. Blaine and Tracey smiled. They loved how the content of President Lincoln's presentation was data they had learned and the images were all pictures they had taken.

"Next are flowering plants," Cecil read, overflowing with happiness. "Flowering plants reproduce with flowers. These flowers are pollinated by the wind or by insects. Once pollinated, they develop into seeds that are often encased by a protective fruit."

A beautiful red rose in a vase was displayed on the screen with the text. This picture had been taken at the castle in Scotland where the twins, along with their expert Fiona McRay, had been framed as jewel thieves.

President Lincoln hit the keyboard, and the image changed to a giant ombú tree, which the twins recalled wasn't really a tree at all, but actually an herb. The twins had been tied to this plant by some rascally wranglers when they had joined the gaucho with ten names on his quest to find the man who killed his gray fox.

Cecil read the text on this new image. "Grasses and herbs are also considered to be flowering plants. Grasses have an extensive root system that helps to reduce erosion. Herbs are plants that are used for flavor, medicine, food, and perfume."

The lab assistant clicked once more, leading the scientist to read on. "Next, we have carnivorous plants," Cecil said. "These plants get the nutrients they need from insects or other small invertebrates that get caught in their traps. They are considered flowering plants as well."

There was another click by the prairie dog. "Then, we have parasitic plants. These are plants that grow on other plants. They get the nutrients they need to survive from their host."

There had been back-to-back images of pictures the twins had taken in Borneo. An image of a pitcher plant to represent carnivorous plants and an image of a giant rafflesia flower, representing parasitic plants. Borneo had been one of the most beautiful locations so far, and being chased by pirates while there had also made it one of the craziest adventures of the summer.

"Last, but not least, we have fungi and molds!" Cecil joyfully bellowed. "Fungi and molds are not really plants. Instead, they make up their own Kingdom of living things."

The last picture of the presentation was one Tracey had taken in Peru of some shelf fungus growing on a tree stump. She had snapped the photo right after she had helped stop a battle from breaking out between a native Amazonian tribe and some archaeologists and loggers.

Upon completion of the presentation, President Lincoln took a bow while Uncle Cecil clapped his hands in a wide circular motion in front of his body.

"What are you doing, Uncle Cecil?" Blaine asked.

"I'm giving Linc-dog a round of applause! Get it? Round… of applause?"

Blaine and Tracey just shook their heads. They got their uncle's joke. They just didn't think it was all that funny.

Cecil clapped in this manner for a little while longer. Then, all at once, he stopped and quickly pulled out a smartphone from each of the side pockets on his white lab coat and pointed them at the twins, almost like he was drawing two pistols for an old western standoff.

"Are you two ready for a new subject of science?" he asked, with a cowboy-ish voice and a stern face that lasted about one second before it cracked a smile.

The twins nodded, grinning from ear to ear.

Cecil tossed the two phones to the children as he said, "Train, Blaisey, here are your new and improved smartphones! They are a bit bulkier because President Lincoln and I added a new waterproof casing. After Blaine dropped his phone in the ocean while wrestling with the Man With No Eyebrows, and then experienced the subsequent zip-a-roo around the world with random caribiner coordinates, we decided to make it absolutely certain that your phones could handle being fully submerged!"

Blaine shivered a bit as he remembered the harrowing incident his uncle mentioned.

"We also put the finishing touches on the new application we were talking about yesterday," Cecil continued. "But before I brief you on that, let's review how all the phone's other applications work."

The twelve-year-olds nodded again.

"First, we have the LINLOC application, which as you know stands for Line Locations. This application gives you the precise longitude and latitude coordinates to the place in the world you will be zipping. It also lists the topics you are tasked with studying and the name of a local expert who will help facilitate your learning. As always, President Lincoln and I will be able to observe your progress down here in the basement on the tracking screen."

The Sassafras twins' heads swiveled to look at the tracking screen, which was the same screen on which they had just viewed President Lincoln's brief presentation. When the screen was on tracking mode, two little green lights representing the twins could be seen wherever they were in the world at the time. This way, Cecil always knew his niece's and nephew's locations. This screen could also function as a "data screen," where Uncle Cecil could view all the data the children had texted in while on their journey.

"The next application, as you know, is the SCIDAT app," Cecil said, swirling his hands around as he talked. He was the kind of person who was unable to remain still for even a second. "SCIDAT stands for scientific data. This application is used to log all the required data for each topic you will be studying. Once logged, just push send, and then all the data will be visible to me down here on the screen."

All the information Cecil was giving them was review for the twins, but they didn't mind at all. They loved talking about anything having to do with their summer science adventure!

"Then, there is the Archive app and the Microscope app," the eldest Sassafras continued. "Along with your phone's high resolution cameras, these two applications help you two little Sassa-ma-frasses get all the pictures you need to text in with your SCIDAT data."

Cecil paused, as he took a long smiling deep breath. Then, he slowly stretched his long arms out as high and wide as they would go—as if he was about to reveal something hugely exciting.

"And now for your new application!" he finally announced. "Behold, the Taser application!"

All four of the twins' eyes widened at this. "The Taser application?" Blaine and Tracey questioned, in excited unison.

"Yes, the Taser application!" their uncle confirmed. "The

two of you have lived through some amazingful adventures this summer and, in the process, have overcome some scarifying obstacles. The biggest of which very well may be this mysterious Man with No Eyebrows you have mentioned. For unknown reasons, he seems determined to derail you two from the track toward scientific knowledge. But now, with this brand-spanking-new Taser application, you two can give him a good zap! Simply open up the Taser application, place the top end of your phone against the desired target, and press the 'TASER NOW' button. At that moment, fifty thousand volts of electricity will shoot any joker or jerk with a not-so-jolly jolt. This Taser will not harm or injure anyone, but it will definitely incapacitate them for a couple of minutes."

Cecil paused and gave a chuckle/frown combo. The chuckling was normal, but the frown was not.

"Over the past few weeks, I have come to the realization that the Man with No Eyebrows is a serious threat," he said. "And President Lincoln and I thought a serious threat deserved a serious application. So that is the reason for the new app, Train and Blaisey."

The twins held their smartphones in their hands and looked at them with a new respect and awe. These little puppies were capable of so much, and now they packed a serious punch. Tracey hoped she never had to use this newly unveiled Taser application. Blaine, on the other hand, was ready to try it out right now. He wondered what it would feel like if he zapped himself.

"Now, to wrap up our little zip-tastic science-mazing review here," Cecil effectively cut off his adolescent nephew's thought process. "This glitch-free, invisible zip-line, smartphone-connected mode of globetrotting is designed to drop you as close as possible to your local experts without being detected. The goal is for you to learn science. As always, you can call me if you need me, but remember: it is very important to keep the existence of these wonderiffic zip-lines a secret."

A pause by Uncle Cecil and a hiccup from President Lincoln passed before the scientist continued. "And now, are the two of you ready to zip away for the start of your Earth science leg of learning? You'll go to several fabtastic locations around the world and study fantabulous topics such as climates, weather, atmosphere, natural cycles, oceans, rivers, and more!"

"YES!" the Sassafras twins shouted in unison.

"Great," Cecil shouted. "Then let's hop to it . . . right after we go back and pay for our groceries at the Left-Handed Turtle."

Blaine and Tracey stood there with arms raised and fading smiles on their faces. They felt somewhat gypped; Uncle Cecil had made it sound like they were going to zip away at this moment, but now it looked like they had to wait a little longer.

Oh, well, they thought. They were still the two luckiest kids on the planet to be able to get to have this kind of adventure at all.

Uncle Cecil went to retrieve his left shoe, and soon the trio found themselves on the sidewalk headed back to the market. The going was slow, as they again cautiously moved remaining on the lookout for the 'guardian beast.'

"Well hello there, Cecil," Mrs. Pascapali, Cecil's next door neighbor said kindly from her front porch, waving her hand. "I see you still have your niece and nephew visiting with you."

"I sure do," Cecil stated with a smile. "Train and Blaisey will be here all summer!"

"That's wonderful!" Mrs. Pascapali said, smiling and nodding her head.

"Ruff, ruff!"

"What's that, Mrs. Pascapali?" Cecil asked, confused at the last thing his neighbor had said.

"Ruff, ruff!" the scientist heard again.

"Mrs. Pascapali, are you barking?" he asked.

"It's not Mrs. Pascapali," Tracey said, pulling on her uncle's white lab coat. "It's Old Man Grusher's dog, and he's coming this way!"

"RUN! RUN LIKE THE WIND!" Cecil screamed at the top of his lungs.

The three Sassafrases scattered in three different directions.

Tracey ran across the street, Blaine ran up into Mrs. Pascapali's yard, and Cecil skittered straight up the sidewalk, screaming like a young child getting a splinter pulled. The guardian beast yapped at all three of the running Sassafrases, but in the end, of course, chose Cecil to chase. This caused the redheaded scientist to run and scream all the more fervently.

Tracey, realizing it was no big deal, slowed her pace a bit, staying safely on the opposite side of the street while heading in the same direction as her uncle and the dog.

Blaine, however, hopped the short fence separating Mrs. Pascapali's side yard from the house at 1108. He stopped and hid behind a bush at the side of the house, panting ever so slightly.

"Wait, what in the world am I scared of?" Blaine abruptly thought to himself. "I'm not afraid of Old Man Grusher's poodle. Why am I hiding?" Uncle Cecil's screams had served to transfer enough fear for Blaine to seek out the cover of the bush, but reality quickly set in. The boy had to laugh at himself as he stood up from his hiding place. He started to step out from behind the bush, when suddenly he felt something hard crash against the back of his head. Everything went black.

## Chapter 2: O-o-o-o-klahoma

### *Where the wind comes sweeping down the plains*

"Welcome to the Left-Handed Turtle! Welcome to the Left-Handed Turtle!"

"Preston! Hurry, hurry, and close the front door of the store! A beast is chasing us!"

"Mr. Sassafras, the doors are already closed. They're automatic."

"Oh, good. Blaisey, I see you made it here without getting eaten. That is great. Where is Train? Is he not here? Oh, fiddle-sox! The guardian beast must have gotten him! Oh, no, this is terri-bad-full!"

"Uncle Cecil, it's okay. Let's just calm down. Take a nice long deep breath," Tracey reassured.

The frantic scientist did as his niece suggested, and after a few moments of normal breathing and realizing he was no longer being chased, Cecil settled back into his normal self. He placed his left shoe up on the counter and proceeded to pull money out of it to pay for his groceries.

"What was the total again, Preston?" Cecil asked the Left-Handed store clerk.

"It was three hundred seventy-six dollars and forty-two cents, sir," Preston answered respectfully, but with a crackly voice.

The scientist paid the correct sum, and then he and Tracey, with the help of the clerk, carried the bags outside. "Thanks, Preston. You've done the Left-Handed Turtle proud. Now will you go inside and call me a cab?"

"A cab, sir?"

"Yes, Preston. My niece, our groceries, and I are going to take a taxi cab back home. That way we won't have to cross paths with the guardian beast again."

"Yes sir," Preston responded, and then he quickly turned, went back into the store, and immediately did as the patron had asked.

"Upon calculated deduction," Cecil said, as he and Tracey sat on a yellow parking block, waiting for their cab, "the likely reason for Blaine's absence is not that he got eaten by Old Man Grusher's dog, but rather that he got separated from us and figured that he would go ahead and start a new scientific subject. I bet he put on his harness, found the coordinates on the LINLOC application, and went ahead and zipped to the first earth science location to meet you!"

"Do you really think so?" Tracey asked.

"I certainly do," Cecil responded. "So, right this moment, the thing I think you should do is zip off to meet him. Do you have your harness, three-ringed carabiner, and smartphone with you?"

"Yes I do, Uncle Cecil. They're right here in my backpack."

"Then, by all means, let's move forward with this plan. Go ahead and get your phone out and open up LINLOC to see what

## Volume 4: Earth Science

your first location is!"

"But what about all these groceries? Don't you need help getting them home?"

"Nonsense. I will manage fine. The taxi will get me safely past the guardian beast, and then President Lincoln can help me get the groceries inside. You go on, Blaisey. It's time to zip!"

It was exciting, but Tracey wasn't sure that Blaine had really zipped off without her. Yes, he did tend to be a little more spontaneous than she was, but he wasn't completely irresponsible. Blaine did joke around with her often, probably because he viewed himself as her older brother, even though he was only older by five minutes and fourteen seconds. But taking off on the zip-lines and leaving her behind wasn't something he had ever done on purpose, and she couldn't picture him doing it now. As Tracey opened the LINLOC app, she had a nagging suspicion that something was wrong.

"Well, what does it say?" Cecil asked, with a sincerely excited smile on his face.

"Oklahoma City, Oklahoma," the girl read. "Longitude 35° 32' 08", latitude -97° 28' 59". The local expert's name is Sylvia Thunderstone, and the topics for study are wind, global wind patterns, downbursts, and tornadoes."

"Oh, hooray! Hooray hippity hippity hoola-hoop hooray!" Cecil exclaimed in total exhilaration. He was beyond excited that his niece and nephew were starting a new scientific zip-line adventure.

The Sassafras Science Adventures

"Blaisey, go join your brother. Have a fantastic time studying earth science face-to-face! I'll see the two of you when you get back!"

Tracey nodded, smiling. She was excited, but the girl was still a little wary of riding the zip-lines without her twin brother. Nevertheless, she put on her helmet and harness, turned the rings on her carabiner to the correct longitude and latitude coordinates, and let the carabiner snap shut. As was always the case, when the carabiner snapped shut, it would automatically find and connect to the correct invisible zip-line. When this happened, it pulled Tracey up into the air a few feet, which Tracey was accustomed to, but she could only imagine what someone would think if they saw her hovering a few feet off the ground now, here in the parking lot of the Left-Handed Turtle Neighborhood Market.

Tracey knew she now had approximately seven seconds before she zipped off at the speed of light to Oklahoma City. Uncle Cecil gave one last smile, accompanied with a wave and then, Whoosh! Tracey was zipping through swirls of light.

Even though she was upset that her brother wasn't with her, the Sassafras girl couldn't help but crack a huge, ecstatic, fully sincere smile. There was absolutely nothing that could compare to what she was experiencing right now. Invisible zip-line travel was the best!

The girl's travel ended with a jerk. Her carabiner automatically unclipped from the line, and then her body slumped down, sapped of strength and sight. All of this was customary, and it didn't alarm Tracey because she knew this was how the lines were supposed to work. She knew that, rather quickly, her sight and strength would return, which they did.

The first thing Tracey did with her restored sight was look around her landing spot for Blaine, but he was nowhere to be seen. "Where have I landed?" Tracey thought to herself. "In a tent? No, there is a wooden floor, but the arched ceiling is low and made of cloth. What is this place? Look at all those cowboy hats and bullwhips and bandanas. Hey wait! This place looks familiar. I'm pretty sure—"

Tracey's train of thought was interrupted by the sound of a harmonica, accompanied by a masculine cowboy-ish singing voice.

"O-o-o-o-klahoma, where the wind comes sweeping down the plain. And the wavin' wheat can sure smell sweet when the wind comes right behind the rain…"

The song continued on, and as it did, Tracey began to think that just like her landing spot, that voice was familiar too. The Sassafras girl crawled toward the opening she noticed and peeked her head out—what Tracey saw made her smile! There, sitting in front of a small crowd singing and playing a harmonica was none other than Sylvester "Doc" Hibbel. He had been the twins' local expert in the United Arab Emirates, where they had studied the integumentary system of the human body. Now, Tracey knew where she'd landed! In Doc Hibbel's covered wagon—the same wagon that Blaine and Tracey had ridden in during the Wind Tower 100 across the Arabian Desert. Strangely, the covered wagon was parked inside what looked like the wide halls of a museum.

The Doc was dressed like a cowboy, and the small crowd he was singing to seemed to really enjoy his performance. He finished his song with a little harmonica solo, and then he addressed his audience.

"I'm so glad that y'all have visited us here today at the Cowboy Hall of Fame in Oklahoma City, Oklahoma. Some folks call me Doc. My mother named my Sylvester Robert Hibbel. My wife calls me lazy or boring. But you can call me friend. I have the privilege of serving as a trustee here at the Hall."

"Oh, yeah," Tracey said to herself, as she remembered that in the UAE, the Doc had told them he was from Oklahoma City and that he was affiliated with the Cowboy Hall of Fame.

"We have something extra special we want to show y'all today," Hibbel continued. "It goes right along with the words to the first line of the song I just attempted to sing. Does anybody remember how the song started?"

A little boy raised his hand and then shouted out the answer, "Oklahoma, where the wind comes sweeping down the plain!"

"That's right, son," Doc Hibbel confirmed and repeated. "'Oklahoma, where the wind comes sweeping down the plain.' Now, I would love to stand here and keep yapping my trap, but instead, I will pass the bean can to someone more interesting than me. Cowboys and Cowgirls, I present to you Dr. Sylvia Thunderstone and her stellar steed, Lucille!"

Hibbel gestured toward the end of the hallway, where a set of large doors opened up and in drove the strangest looking vehicle Tracey had ever seen. She couldn't tell if it was a car, an R.V., a tank, or a combination of all three. It rolled slowly and carefully up to where Doc Hibbel and the small anticipatory crowd were. The vehicle parked and then nothing happened for a few seconds, giving the excited onlookers a chance to gawk over the detail and design of this peculiarly awesome car.

It had six wheels. It looked to be armored, and it had a window built into its roof. All at once, a V-wing door on the side opened up and out stepped a strikingly beautiful lady with brunette hair, big colorful eyes, and a kind smile on her face.

"Oklahoma, where the wind comes sweeping down the plains is right, y'all," she said to those gathered. "It's so nice to meet ya today. My name is Sylvia Thunderstone. I am a meteorologist fresh out of my doctoral courses, and I am here today to talk to y'all a little bit about that Oklahoma wind, as well as introduce you to Lucille here."

"Lucille?" questioned the same boy who had answered Doc Hibbel out loud earlier.

"Yes, Lucille," Sylvia confirmed, as she patted the strange vehicle with her hand.

"Lucille is being presented for the first time today to y'all at the Cowboy Hall of Fame. She is the very latest, cutting-edge, state-of-the-art, technologically-advanced, meteorologically-superior, storm-chasing-vehicle to be designed and assembled!" Sylvia announced proudly, as the small crowd gasped in wonder.

"We call her Lucille, after Lucille Mulhall, who was from Oklahoma. She was the very first show cowgirl, known as the 'Queen of the Western Prairie.' She performed in many rodeos and Wild West shows throughout her career." Thunderstone paused as she thought about Lucille Mulhall, one of her heroes.

She smiled and patted Lucille, before continuing. "Our hope for Lucille is that she will be a hero used to save thousands of lives!" Sylvia continued. "She'll do this by helping us to really figure out how dangerous storms work."

"I-yip-I-yo-ee-yay!" Doc Hibbel exclaimed, unable to conceal his excitement over this brand new storm-chasing car.

Sylvia smiled at the Cowboy Hall of Fame trustee and then got to the subject at hand. "Now, to address that wind that so often comes sweeping down the plains and across the prairies here in Oklahoma. Wind is actually the movement of atmospheric gases on a large scale. In other words, it's the movement of air. And since wind is something that you can't actually see, we describe it by using two factors—speed and direction." Many people were now nodding their heads, understanding what Thunderstone was saying.

"Wind is caused by the uneven heating of the surface of the earth," Sylvia continued. "The earth's surface, of course, is a mixture of land and water. Land and water both absorb heat from the sun's rays at different rates. During the day, the sun heats up the surface of the earth and the air around it. The air over land heats up faster than the air over water. Plus, we have the air over land that receives direct sunlight, which heats up faster than the air over land that receives indirect sunlight. Since the warm air weighs less, it rises, causing a change in air pressure. The cool air moves in to replace the space where the warm air was, and this movement of warm and cool air causes wind. At night, this happens in reverse. The air over land cools more quickly than the air over water, and wind is created once more."

The crowd leaned in a bit, interested to hear what Dr. Thunderstone had to say.

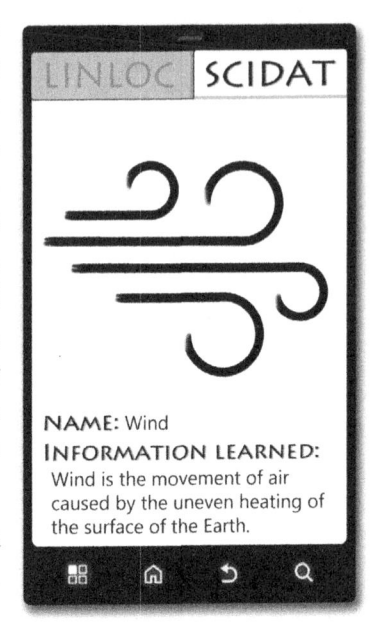

**NAME:** Wind
**INFORMATION LEARNED:** Wind is the movement of air caused by the uneven heating of the surface of the Earth.

"When there's lots of wind, we can harness its power and turn it into energy that we can use. This is known as wind power, which we use to generate electricity these days. But in the past, it was used to pump water out on the prairies."

Sylvia paused here and smiled. "Maybe you remember there was another few lines in the song Old Doc Hibbel was singing that said:

'Gonna bring you barley, carrots and potaters,

Pasture for the cattle, spinach and tomaters,

Flowers on the prairie where the Junebugs zoom,

Plenty of air and plenty of room.'"

"I remember!" The outgoing boy called out.

"Well, Oklahoma is mostly prairie, which is a type of temperate grassland. Now, there are two types of grasslands: temperate and tropical. The savannah, which is a tropical grassland, for instance, has a hot wet season that lasts for a few months and a slightly cooler dry season that lasts for about eight months. But here in the temperate prairie, we have cold winters and warm summers, just like the steppes of Europe and the pampas of South America."

Tracey had to smile at the pampas reference. She and her brother had had quite the adventure in the Argentinian pampas.

"The temperature on the prairie can be as low as negative twenty degrees Fahrenheit in the winter and as high as a hundred degrees in the summer," Thunderstone said. "But the averages are around twenty degrees in January and around seventy in July. The average rainfall is between ten and thirty inches, but most of that occurs in the summer months. Wind storms are pretty common here on the prairie. These storms are basically periods of high winds and strong gusts with no rain."

Upon the conclusion of the lovely Dr. Thunderstone's talk, the small crowd applauded. Tracey, who was still in the back of the covered wagon, opened up the archive app on her phone and started looking for a suitable image to represent wind. She had enjoyed hearing from Sylvia, and it was good to see Old Doc Hibbel again, but Tracey really did not want to be entering SCIDAT data and pictures alone. Where in the world was Blaine?

### *Easterlies, Westerlies, and Forgotten Memories*

Where in the world was he? Blaine slowly opened his eyes and tried to get up from the ground where he was laying. All at once, blood rushed to his head and he felt a throbbing pain at the back of his skull. He winced and immediately fell back down on his stomach. From this position, he slowly craned his aching head around, trying to figure out where he was. The last thing he remembered was running from Old Man Grusher's poodle into Mrs. Pascapali's yard. He couldn't remember anything after that.

He had been outside, in the yard—but now, by the looks of it, he was inside. "But inside where?" he wondered.

Maybe a basement? Was he in Uncle Cecil's basement?

No, no, that couldn't be. This place was way too clean and organized to be his uncle's basement.

Suddenly, a voice that terrorized his heart and made his skin crawl called out his name. "Blaine Sassafras," the voice said, sounding like a score of serpents slithering over sand.

Blaine froze. Who was standing behind him? He had never heard a voice so creepy. The hair stood up on the back of his neck.

"I see you are too afraid to turn around," the voice slithered again. "So let me walk around where you can see me."

Heavy, but slow, boot thuds sounded off of the floor as the owner of the voice walked from behind Blaine to a spot in front of him. The Sassafras boy could now only see the voice's boots, but he was too scared to look up and see who the boots' owner was.

"Look up at me, you little Trash-a-fras! Look me in the face!" the voice boomed, almost causing Blaine to go pulse-less.

The twelve-year-old took a dry-throated gulp and obeyed the command, slowly and timidly. When Blaine's gaze finally reached the voice's face, shock enveloped him. He was staring into the face of the Man with No Eyebrows!

"Surprised, are you?" the eyebrowless man said with a sinister smile. He was obviously enjoying the fact that Blaine was so surprised.

"I will get straight to the point, my little friend," the man said, with an obvious sarcastic use of the word 'friend'. "I am here to destroy your zip-lining adventure!"

That was not what Blaine wanted to hear.

The man continued. "Do you see my face? Do you notice how I don't have any eyebrows? Do you see? Do you see boy?"

Blaine nodded reluctantly.

"Of course you see!" That is why you and your twin sister call me the Man with No Eyebrows, isn't it?"

Blaine kept nodding, as he tried to control the fear that continued to well up in him.

"It is your uncle's fault that I don't have any eyebrows! That despicable man named Cecil Sassafras is the culprit! He is the one who took my eyebrows from me! I have vowed my revenge on him. I have promised myself that I will destroy anything that he creates, because of the grief that he has caused me. Including his greatest invention to date—the invisible zip-lines. So as of late, I have put all of my vengeful energy into stopping you and Tracey from traveling around the globe on these zip—lines, learning science face-to-face."

Even in the midst of his fear, Blaine was actually experiencing some relief. He now had an answer to one of his biggest questions. Now, he knew the motivation of the Man with No Eyebrows! He and Tracey had been seeking this answer all summer long.

"Yes, I want to destroy your adventure," Mr. No Eyebrows continued in his heartless voice. "And by destroying that, I will also be destroying Cecil Sassafras. Something I have been trying to do all summer, but you two little Trash-a-frasses have proven to be more resilient than I would have thought possible."

Blaine actually cracked a small proud smile at that statement, which the Man with No Eyebrows saw, so he raised his voice even more.

"But your resiliency has now come to an end!" he boomed. "Because you, my friend, are about to face the 'Forget-O-Nator!'"

"The 'Forget-O-Nator?' Blaine tried to ask out loud, but because of his dry throat, it remained a silent, panicked question in his mind.

The Man with No Eyebrows took a few steps over to the center of his basement and then stopped again. Blaine's fearful gaze followed the path of the intimidating antagonist, and the boy soon found he was staring at something that looked a little like a port-a-potty.

The man boomed. "I have watched you on tiny cameras I have hidden around your uncle's house. I have stolen a jeep, leaving you stranded out among wild animals. I have barricaded you inside of an ancient tomb. I cut down a tree that you were in the top of. I have spied on you with a robot hummingbird. I have sabotaged your uncle's computer and jumbled up your scientific data. I have trapped you inside indestructible boxes. I have chased you with robot squirrels. And I have stalked you all over the globe while wearing the magically disappearing Dark Cape suit."

Blaine was still too scared to talk, but his brain was firing on all cylinders. Pieces of the puzzle were being put together as questions were being answered.

"I did all of this with the sole purpose of stopping you and your sister from learning science, which would therefore crush Cecil Sassafras's hopes and dreams. But none of those plans succeeded. You and your sister just keep zipping and zipping and zipping to new locations. And you just keep learning and learning and learning science."

The more the Man with No Eyebrows spoke, the angrier he became. His teeth clenched and his fists tightened. Blaine, however, was relaxing with time because the man was reminding him of all the times they had bested this eyebrowless villain. Blaine hoped this time would be no different.

"But nothing I just spoke of matters anyway," the Man with No Eyebrows continued. "Nothing you have seen or experienced this summer will stay in your brain after a few moments in this machine!"

The Man pointed at the port-o-potty look-alike. "Welcome to the Forget-O-Nator! I just open the door and put you inside. I lock the door. Then I turn the one and only knob to full capacity.

Two minutes later, out you come, voila, your memory is wiped completely clean! Nothing you have ever done or learned will remain. You will be forget-o-nated. I will be avenged! And Cecil Sassafras will be destroyed!"

The Man with No Eyebrows lifted his hands into the air and let out the scariest laugh Blaine had ever heard. Then, before the Sassafras twin knew what was happening, the Man with No Eyebrows had lifted him off the ground, thrown him into the Forget-O-Nator machine, and locked him inside. Blaine could hear the eyebrowless enemy laughing on the outside. He could also hear a clicking noise as the wicked man turned the one and only knob, cranking the machine to life.

Fluorescent lights on the inside of the Forget-O-Nator immediately began to flicker on brighter and brighter. Blaine could almost feel the memories and knowledge in his brain being tugged out. As fast as humanly possible, Blaine did the only thing he could think of to do. He pulled out his smartphone and opened up the new Taser application. He reached out, touched the wall of the Forget-O-Nator, and pressed the "TASER NOW" button. The last thing the boy remembered was the sound of an explosion and a shower of sparks.

"You haven't made it home yet?"

"I sure haven't, Trace-a-ma-fras. When the taxi pulled up in front of 1104 to drop me off, the guardian beast was standing on the sidewalk right in front of the house. So, I am currently having the taxi driver take me around the block repeatedly until the cruel beast leaves."

"But Uncle Cecil, you just have to get back home and check

the tracking screen to see where Blaine is. I have checked all over the Cowboy Hall of Fame, and he is not here!"

"You are right, Blaisey. You are right—easy-peezy, kinda-sneezy. Okay then, okie-dokie, okie-hokie—I'm gonna do it. I'm gonna get back home with the groceries. I'm gonna get past the guardian beast. And I'm gonna get back down to the basement to check the tracking screen for you. I'm gonna do it. And I will call you back just as soon as I find out where your brother is, okay?"

"Okay," Tracey replied softly. Then she disconnected the call with her uncle.

"Tracey Sassafras?" the girl immediately heard. "Tracey Sassafras, is that you?"

Tracey turned around and saw that it was Doc Hibbel who had called her name, and he was walking her way, accompanied by Dr. Sylvia Thunderstone. "Well, it sure is you, little lady! How in tarnation have you been? Where is your brother? Is he here with you?"

"Oh, he's around somewhere," Tracey said sheepishly.

"Tracey, this is Dr. Sylvia Thunderstone," Hibbel introduced. "And, doctor, this is Tracey Sassafras. She and her brother rode with me in my covered wagon during the Wind Tower 100 just a bit back when I was in the UAE."

Thunderstone smiled and looked impressed as she voiced, "Oh, wow! Nice to meet you, Tracey! Sylvester told me about some of your adventures out there in the Arabian Desert."

"Yes, it was quite the adventure," Tracey laughed. "I just heard your talk about wind, Dr. Thunderstone, and watched you introduce Lucille—both topics were fascinating!"

"Oh, thanks, Tracey," the doctor responded. "But you can just call me Sylvia."

"Oh, no, no, no, you can't just leave it at that," Old Doc

Hibbel interjected. "Tracey, you just have to hear the story behind Dr. Thunderstone's name!"

Sylvia laughed and rolled her eyes in a good-mannered way, knowing she wasn't about to stop the Doc from telling what he considered to be a good story.

"Sylvia's great, great grandfather was a full-blooded Native American from the Shawnee tribe. His name was Moonbeam," Hibbel started the tale with enthusiasm. "One particular summer day, out on the prairie where the Shawnee lived, a horrendous storm flared up and came rolling directly toward where the tribe's teepees were set up. It was the worst storm any in the tribe had ever seen, with strong winds, dark ominous clouds, hailstones the size of grapefruit, unpredictable flashing lightning bolts, and loud booming thunder that seemed to literally shake the ground.

"All the members of the tribe ran in utter fear for any shelter they could find—all except one. Moonbeam set his face like stone, turned toward the storm, and ran directly into it. Several of his tribesmates screamed his name, crying for him to come back and hide with them, and yelling that he was only a boy. They shouted that the lightning and thunder were going to seek him out and bring him death. 'I must rescue the horses,' they heard him shout back just before his silhouette was engulfed by a black sheet of rain pellets and hailstones as the storm thundered over and around the tribe for several hours, battering them to the brink of hopelessness.

"Finally it passed, and everyone's first thoughts were, 'Where is Moonbeam? Did he survive?' They ran out into the prairie where their teepees and horses had been. What they found was that their homes had all been destroyed or blown away. But what truly astounded them that day was a brazen figure huddled down in a shallow gully with an entire herd of shivering horses. It was Moonbeam. He was beaten and bruised and soaking wet, but he was very much alive, and he had saved every one of the tribe's horses.

"That night, around the fire, the chief of the tribe changed Moonbeam's name to Thunder Stone, for the boy who set his face like stone, ran into the thunderstorm, and saved the horses," Doc Hibbel concluded.

Tracey wanted to say "Wow," but she was left speechless by the truly amazing story.

"Obviously, Sylvia derives her name from her great, great grandfather," Doc Hibbel said, as they began walking together down the wide hall. "But she also initially became interested as a child in meteorology because of this story."

Sylvia nodded, confirming this was true.

"And here she is now with a doctorate in meteorology, still willing to come out here to the Cowboy Hall of Fame and teach some science to us mooncalves, as well as let us meet Lucille, the latest and greatest storm chasing car! She is a great Oklahoman, and her great, great grandfather was a great Oklahoman too!"

All the Doc's compliments caused Sylvia to blush a bit.

"Look here, Tracey. Look at this!" Doc Hibbel said this as he pointed to a large painting on the wall. "Here he is, right here!" The image in the painting was the strong face of a Native American man facing a ferocious looking storm; the painting was entitled 'Thunderstone.'

"This is him?" Tracey asked, excited. "This is your great, great grandfather?"

Sylvia nodded proudly.

"Wow!" Tracey exclaimed. This time she was able to get the words out of her mouth. "Dr. Thunderstone…I mean Sylvia…I have really enjoyed hearing your family's story, and as I said before, I enjoyed your talk about wind. I would love to hear anything else you would like to share about the science of weather."

Thunderstone smiled, very thankful for the attention and

compliments. But not wanting to keep the focus on herself, she gladly changed the subject. "Yes, you know Tracey, when I was talking about wind earlier, I also meant to add a little about global wind patterns. Would you like to hear about that?"

Tracey nodded yes.

"Okay, then," Sylvia said. "Well, the movement of air around the globe is known as global wind patterns. On a large scale, the winds that circle the earth are created because the land at the equator is heated more than the land at the poles. Another factor that affects the winds around the globe is the spinning motion of the earth. This is known as the Coriolis Effect."

As Sylvia talked, her hands demonstrated the Earth's movements while the three continued to walk down the wide halls of the museum. Out of the corner of her eye Tracey saw more amazing paintings, detailed statues, and different kinds of memorabilia commemorating famous cowboys and great figures in Wild West history.

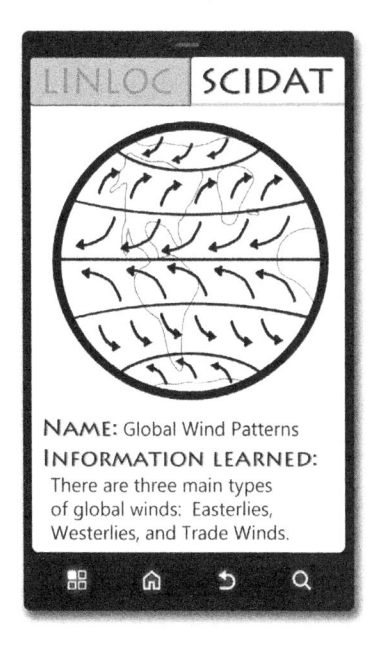

"There are three main types of global winds: Easterlies, Westerlies, and Trade Winds," Thunderstone continued. "Easterlies, also known as 'Polar Easterlies,' are winds that are found near the north and south poles. They blow up to the poles and curve from east to west. Westerlies, sometimes referred to as 'Prevailing Westerlies,' are winds that are found in between the equator and the poles. They blow slightly toward the poles from west to east. And Trade Winds are winds found near the

**NAME:** Global Wind Patterns
**INFORMATION LEARNED:** There are three main types of global winds: Easterlies, Westerlies, and Trade Winds.

equator. They flow north or south toward the equator and curve west due to the spin of the earth."

Sylvia let her hands fall away from the imaginary globe she had been using in the air to demonstrate the flow of the global wind patterns before she continued. "And then there are the jet streams—one of the coolest facets of the global wind patterns, in my opinion. Jet streams are rivers of fast moving air about five to nine miles above the earth's surface. They form at the boundaries of where the polar and temperate, or tropical, air meet. Because of the effect of the rotation of the earth, jet streams flow from west to east in a wave-like manner. Isn't that cool?"

Tracey nodded her head and was about to comment on the subject of jet streams, when the sound of a loud piercing siren suddenly filled the air, resounding up and down the corridors of the Cowboy Hall of Fame.

"What in the world is that?" Tracey asked. "A burglar alarm?"

"No. It's much worse!" Dr. Sylvia Thunderstone shouted. "That's the tornado siren!"

## Chapter 3: Lucille's First Rodeo
### *Damaging Downbursts*

Cecil rubbed his eyes and looked again. "Is that really him?" he asked himself. The figure he was looking at continued to wander down the sidewalk until the individual eventually stopped directly in front of his house.

"It is him! It's Train!" the scientist exclaimed to himself.

Cecil, who had just gotten out of his taxi and was standing on his front porch, surrounded by blue grocery sacks, ran down off the porch toward his confused looking nephew with outstretched arms.

"Train! My calculated deductions were wrong! You didn't zip away without your sister. Blaisey's inclinations were right! Are you okay? You look a little woozy. Where have you been? Are you okay? Are you okay, Train?"

The twelve-year-old boy looked up with dazed eyes.

"Yes, I'm okay, sir," he said groggily.

"Sir?" Cecil questioned.

Blaine shook his head and searched for the correct name. "I mean uncle....Sal? Sam? Skip? Cicero? Cecil! Oh, yeah, Cecil!"

A look of concern formed on the scientist's face. What had happened to his nephew?

He got the boy safely into the house and down to the basement. Blaine took a seat while Cecil took his nephew's backpack and checked its contents, pulling out the boy's harness, helmet, three-ringed carabiner, and smartphone. Cecil glanced up at the tracking screen and saw that Tracey was still in Oklahoma City. He looked back down at the contents of the backpack and then back at his nephew's glazed-over face. Was that singed hair over Blaine's ear?

"Where is Tracey?" the boy suddenly asked.

"She is in Oklahoma City, Oklahoma," Cecil replied, pointing at the tracking screen. "That is the first location of your earth science studies. I thought you had zipped away without her, but obviously I was wrong. Jiminy, jiminy, jeeze little weeze, we need to get you to where she is as quickly as possible, but are you okay? What happened to you? Did the guardian beast catch you?"

Blaine looked like he was scanning through his mind for the answer to his uncle's question. "The guardian be…I don't…wait…the dog…Old Man Crusher's miniature poodle was chasing us. You and Tracey ran one way, and I…ran up into Mrs. Polkapalis's yard, then…I…came back here…and met you…right?"

Cecil nodded at his nephew, but he knew that something more than what Blaine just said had to have happened. Maybe the boy was just a little traumatized by being chased by the dog. Of all people, Cecil understood that feeling.

"I am ready," Blaine blurted. "I am ready to open up the LIMELOC app, get the correct coordinates, and zip off to meet Tracey."

The boy's eyes suddenly looked mostly clear again.

"Okie dokie, then," Cecil chuckled, thinking that evidently his nephew was fine after all. "Let's get you harnessed up!"

Black smoke still hovered in the air. Most of the parts and components seemed to still be there, but they were strewn across the floor, broken and burnt.

The Forget-O-Nator had exploded.

How had it happened? What had that boy done to his machine? He traced back through his mind to all that had happened earlier. That boy had run up into his yard, which had been like a gift wrapped with a bow. He had hit the boy on the back of the head with the handle of his shears. He had dragged the boy down to his basement, and when the boy regained consciousness he had revealed his true identity to the boy and had talked gruffly to him. Then he had stuck that boy into the Forget-O-Nator, locked the door, cranked the one and only knob to full capacity, and smiled, thinking everything was working smoothly.

But after fifteen seconds or so, the Forget-O-Nator had suddenly exploded in a shower of sparks and plumes of smoke. Getting hit by flying debris, he himself had been knocked unconscious. When he came to, that boy was gone and his precious machine was in shambles.

The Man with No Eyebrows clenched his teeth in anger and picked himself up off his basement floor where he had been laying.

Again he had been bested by one of those twins! Again he had failed to complete his vengeance on Cecil Sassafras! He threw his head back and let out a roar—a roar that was part defeat, part anger, and part dark determination. He would pick up the pieces that lay across his basement, and he would rebuild the Forget-O-Nator. He was resolved to use this thing on the twins!

"At least that boy was in the machine for fifteen seconds, he thought to himself. Surely those fifteen seconds did something to the boy's brain. Surely some of his knowledge and memories were at least jumbled or pulled at." The man sighed and rubbed his eyebrowless forehead as he optimistically tried to keep his dark heart pessimistic.

"Blaisey. Blaisey!" In the midst of all the commotion caused by the tornado siren, Tracey heard her name being called. Well, at least the name her absent-minded uncle usually called her. "Blaisey! Blaisey, over here!"

The Sassafras girl looked around until she finally spotted the source of the sound. It was Blaine! He was here! And he was in the back of the covered wagon, just like she had been earlier. Had he just landed? What was happening? Tracey rushed over to the wagon.

"Where in the world have you been?" she asked, in a more scolding tone than she would have liked. "And why are you calling me Blaisey?"

"Because that's your name," Blaine answered. "Blaisey…wait no…Tracey…your name is Tracey…You are my sister. Well yes, I guess you already knew that."

Tracey looked at her twin brother with a look of confusion

on her face. Why was he acting and talking so weird?

"Where in the world have you been?" she asked him again.

"All over the world," he said, "zipping around with you—learning science."

"No, I mean today—this morning. Where were you when I zipped here to Oklahoma City by myself? Did Old Man Grusher's dog actually get to you?"

Blaine laughed like that was preposterous. "No! No way could that mini-dog ever catch me. I just…got lost…or something…for a little bit. Look, it doesn't matter. I'm here now, so let's learn some earth science! What's happening here in Okra-homa? What is this place, a museum or something? Why is everyone running around?"

"It's not a museum; it's a Cowboy Hall of Fame. And everyone is running around because a tornado siren just went off. C'mon, get out of the covered wagon! I will introduce you to Sylvia Thunderstone, our new local expert, and we will figure out what to do. Oh, and guess who else is here? You're not going to believe it—Doc Hibbel!"

"Doc who?"

"Doc Hibbel, you know, Sylvester Hibbel. Our local expert back in the United Arab Emirates."

"United Arab what?"

"Never mind. Just get out of the wagon and let's go find Sylvia!"

Blaine clambered down out of the covered wagon and followed his sister as she ran over to a lady standing by herself by one of the museum's emergency exit doors. The pretty lady was looking through the door's window into a stormy sky. Upon the twins' approach, she stopped sky gazing and looked at them.

"Is this your brother, Tracey?" the woman asked.

"It sure is," Tracey confirmed. "Sylvia, this is Blaine. Blaine, this is Dr. Sylvia Thunderstone. She is a meteorologist and a stormchaser."

Blaine shook the meteorologist's hand and then looked through the window out into the sky himself.

"Looks pretty nasty out there, doesn't it?" Sylvia said. "I was just looking closely at the cloud formations to see if we have any wall clouds or derechos."

"Derechos?" Blaine asked. "That sounds like something you would order at a Mexican restaurant."

Thunderstone laughed at Blaine's comment. "Derechos are actually large clusters of strong thunderstorms. They form in a long line and can cause widespread wind damage. The damage is caused by an abundance of downburst winds that derecho storms can produce."

"What is a downburst?" Tracey asked.

"A downburst is a strong downward current of air," the meteorologist answered. "They are caused by rain-cooled air that sinks and rushes back to the ground. When it reaches ground level, it quickly spreads out in all directions, which causes the production of strong damaging winds. Unlike tornados, winds in downbursts blow outward from the point at which the wind hits the land. This forces things in the wind's path out instead of sucking them in like a tornado would.

"These downburst-producing storms typically form

**NAME:** Downburst
**INFORMATION LEARNED:**
A downburst is a strong downward current of air caused by rain-cooled air that sinks and rushes back to the ground.

in late spring and summer. The warm humid air on the grasslands and prairies allows for the grass to grow very tall, but it also sparks lots of storms. On the grasslands, there are no natural barriers like an abundance of trees or mountains, so there is a lot of wind."

Sylvia paused and looked back at the window. "The storm that is forming out there right now does not look to be a downburst-producing storm. It does, however, look to be a tornado-producing storm, which, of course, is why the sirens are going off."

"Sylvia! Sylvia!" The doctor was suddenly cut off by someone calling her name over a cell phone-looking device that was attached to her belt.

"Oh, it's Jayman calling me on the walkie-cell!"

"Jayman?" Tracey asked.

"Walkie-cell?" Blaine asked.

"Yep," Sylvia responded. "Jayman is a friend and colleague of mine who keeps me updated from our weather station. He is also the inventor of this walkie-cell, which has all the tools and functions of an average smartphone but with the added bonus of direct channels for immediate communications like a walkie talkie."

"Sylvia! Sylvia! Come in, Sylvia!" The voice of Jayman crackled over the walkie-cell again.

"I'm here, Jayman. Go ahead," Thunderstone responded.

"Sylvia, we have a confirmed tornado here on the ground!"

"What? Where? Is it near us here at the Cowboy Hall of Fame? How big is it?"

"It's not a very big one yet. Right now, it is registering as an F1. It looks to be about 1.8 miles southeast of all at the C.H.F."

"Are we in its path?"

"No, the twister turned and I think it will miss you guys by at least a mile. But it's probably best for everyone there to take cover, just in case."

"Copy, Jayman. Tornado precautions are currently underway. We will make sure that every last person finds adequate shelter."

"Sylvia," Jayman said in a calmer and possibly even excited tone, "I think this is a great chance to get Lucille into her first rodeo."

Sylvia smiled, knowing exactly what her colleague meant. "I think you're right, Jayman, I'll go get her fired up right now!"

"Okay, you do that, Sylvia, but stay safe. I'll be right here on the other end of this line if you need me. Over and out."

"Over and out, Jayman." Sylvia ended her call with a nervous but excited smile on her face. "Looks like this storm is going to be Lucille's first rodeo! Are you two ready to saddle up?"

"Saddle up? We get to ride Lucille with you?" Tracey asked happily.

"You sure do, if you want to," Sylvia answered. "If you two can handle riding a covered wagon in the Wind Tower 100, I think you can also handle riding Lucille in the wind of Oklahoma."

"Awesome! We're in!" Tracey exclaimed. "Lucille, here we come!"

Both females ran off, but Blaine just stood in his place by the door.

"Wait, what on earth is Lucille?" he asked

"She is the very latest, cutting-edge, state of the art, technologically advanced, meteorologically superior, storm chasing vehicle that Sylvia designed and assembled!" Tracey responded. "C'mon bro—you know you don't want to miss out on this."

Blaine still didn't know exactly what was happening, but he followed the two ladies nonetheless, knowing that during these adventures, questions sometimes had a way of answering themselves.

The three made their way back over to where Lucille was. Sylvia ran up to the tank-looking vehicle and opened up the horizontally hinged side door.

"What is this thing?" Blaine gasped.

"This is Lucille," Tracey said, smiling.

Blaine silently mouthed the word "Awesome" and stepped into the car after his sister. Thunderstone got in last and was about to pull the door down when, suddenly, an out of breath figure stepped out from behind the covered wagon. It was Sylvester Doc Hibbel.

"Whew-wee!" he said with a wry smile. "There are quite a lot of stairs to and from that storm shelter, especially when you run up and down them a couple dozen times!"

"Did you successfully get everyone down there?" Sylvia asked the old cowboy.

"I sure did, Thunderstone. Everyone is safe and accounted for. I pray that no twister touches this place, but if one does, that is a great shelter. Everyone will be fine."

"But what about you, Doc?" Tracey asked with concern. "Why are not down there with them? Are you coming with us?"

Hibbel smiled at Tracey and then gave her a grandfatherly wink. "Don't worry about me, little lady. I'll be fine. I'm gonna let you and your brother take a ride in Lucille here with Dr. Thunderstone by yourselves. Meanwhile, I'm gonna go fetch Ike and Wyatt. I'll tether them up to the wagon, and we'll see if we can find somewhere safe to hunker down."

"But what about the storm?" Tracey asked. "Jayman just said there is a tornado on the ground."

"That's all right, little lady, I've got tornado-proof boots on."

"You do?"

"Sure do," Doc Hibbel said, hiking up his chaps just a little so everyone could see his leather cowboy boots. "These here are 'Old Doc Hibbel's Hand-Stitched Western Boots.' They're tested and approved by Oklahoma cowboys and Indians alike. They are made from the hides of only the meanest and toughest bulls that can be found. On the inside, they fit like cotton and walk like clouds, but on the outside they are gruff and rugged and have an all-around bad attitude. They can kick through cactus, and they can repel the tumbleweed blowing winds of the prairie. They can fend off the coyote's bite, the scorpion's sting, and the rattlesnake's venom. They have been known to outrun stampedes, outjump blazin' bullets, and outlast tornados, whirlwinds, twisters, and whirligigs of all magnitudes. And on top of all that, they come complete with stylish heel-side hidden zipper pockets. Nothing can beat a little tough on your feet—Old Doc Hibbel's Hand-stitched Western Boots."

The twins had to laugh at Sylvester's silly sales pitch. They were reminded of how he was always trying to peddle his western wares with funny little jingles.

"Ride, Lucille, ride!" Doc Hibbel shouted, adding, "giddy up you three! Get this amazing vehicle out into the storm and gather some data that will help us save lives!"

Sylvia gave the Doc a thumbs-up as she pulled down the car's door and locked it shut.

"Do you think he'll be okay?" Tracey asked Sylvia softly.

"Oh, he'll be fine," the doctor replied. "That old cowboy has lived through more tornadoes than anyone I know."

Thunderstone climbed over to the driver's seat and started up the engine.

"Giddy up, indeed," she said. Then she began carefully driving Lucille through the halls of the C.H.F. toward a place where they could exit.

## Thunderstone's Tornado

The twins found that the inside of the storm-chasing vehicle was rather spacious. They couldn't stand all the way up, but there was plenty of room to move around and check out all of the meteorological gadgets. Once the vehicle was outside, the Sassafrases could see and feel just how ominous the weather truly was. There were clouds of all different shades of white, gray, and black swirling around. The wind was blasting at Lucille with major force!

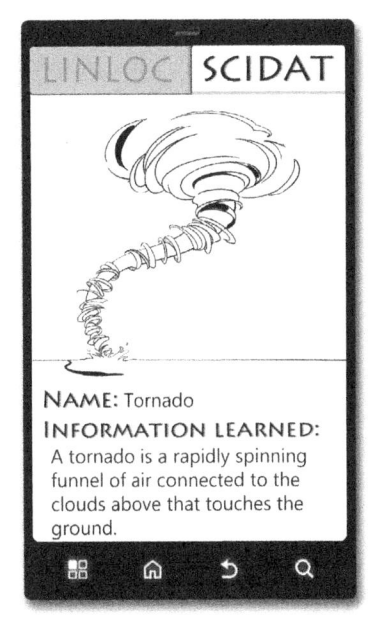

"Tornadoes typically form from strong thunderstorms," Sylvia shared as she drove. "In the midst of these storm clouds, there is hot, humid, fast-moving air swinging upward; and cold, dry air moving downward. These two currents spiral and spin around each other, forming a funnel. If the currents are strong enough, this funnel will reach the ground to form a tornado. By definition, a tornado is a rapidly spinning funnel of air connected to the clouds above that touches the ground. Before a tornado touches the ground, it is known as a funnel cloud. We are still trying to figure out the exact conditions that form a tornado and the exact conditions that cause it to die out."

"And that's why Lucille is here, right? To study and look for those exact conditions?" Blaine offered.

"Right," Sylvia confirmed. "Most tornadoes only last a few minutes, but during that time they can tear up trees and houses and move people, animals, and even cars. So they are very dangerous.

We want to use science and technology to do everything within our power to limit destruction and save lives. "

The twins nodded in whole-hearted agreement at this last statement.

Thunderstone continued. "The Fujitsu scale is used to describe the strength of a tornado. It ranges from F0 to F5 with F5 being the strongest. Each category has a wind speed range and a description of possible damage. The majority of tornadoes spin at around one hundred miles per hour, which in an F1 on the Fujitsu scale. An F1 tornado can cause moderate damage, like snapping trees in half, blowing mobile homes around, and damaging roofs. An F5 tornado can wipe entire neighborhoods away. It can even tear up and blow away pieces of concrete slab foundations."

Now the twins shook their heads in disbelief.

"Most tornadoes form in the spring and occur in an area known as tornado alley, which is made up of the Great Plain States in the United States, including Oklahoma. Over five hundred tornadoes touch down in this area every year."

Just as Sylvia finished giving her information, her walkie-cell crackled to life with Jayman's voice again.

"Sylvia, come in."

"Here I am, Jayman. Reading you loud and clear."

"Sylvia, it looks like our tornado has strengthened from an F1 up to an F2. It has hopped around a bit since I last talked to you, but now according to my radar, it looks to be moving straight up the turnpike toward Tulsa."

"Okay, Jayman, we are headed that direction now. We'll see if we can get some eyes on this F2 and use Lucille to gather some data."

"Sounds good, Sylvia. If you need anything, I'll be on the line. Over and out."

"Over and out, Jayman." Dr. Thunderstone finished the call with her colleague.

While Sylvia drove, she began pulling, turning, and pushing at all kinds of buttons, switches, and levers located all over Lucille's interior. As she did, an array of different gauges and meters lit up, plus several screens turned on, making the inside of this storm-chasing vehicle look like some kind of spaceship. The thing the twins were most impressed with was the button that had folded the interior roof into the car, revealing an entire ceiling of glass.

Sylvia saw Blaine's and Tracey's reactions to the opening of the roof and commented, "Pretty cool, isn't it?"

The twins nodded.

"That glass is four inches thick, but it's still easy to see through. It's the same stuff they use in aquariums to make shark tanks out of."

The Sassafrases gasped and gazed up through the ceiling window into the stormy Oklahoma sky. Ominous clouds were still rolling around overhead, and now they could see that it had started raining.

Sylvia maneuvered the six-wheeled car skillfully through the city streets as the wind continued to blow. Tornado sirens continued to sound. And the rain fell harder and harder. Eventually, they left the smaller streets and drove up onto a wide highway, and just as they did, all three gasped.

There it was—the F2 tornado, spinning wildly just a little north of the highway!

"Whoa," the twins said in awe, as they watched the white funnel swirl from the clouds to the ground, tearing through a field and moving away from them.

"Okay, Blaine and Tracey, here's the plan," Thunderstone informed, full of energy. "We will speed up the turnpike and get ahead of this twister. Then, I will try and park Lucille directly in its path. If we can get this tornado to pass directly over us, we will have the best chance of gathering data!"

"Directly over us?" Blaine questioned. "You mean, we're going to be inside the tornado?"

"That's the plan," Sylvia confirmed.

"But won't it blow us away?" Tracey inquired nervously.

"Not according to our calculations," the storm chaser answered. "Lucille is designed to have wind pass around and over her, allowing little to no wind to get underneath her or to hit her broadside. Additionally, Lucille is equipped with a thirty foot corkscrew anchor. Right now this anchor is coiled up into itself on the bottom side of the car. When we park, I will push this button."

Thunderstone pointed to a green button that said "CORKSCREW" on it. "This will cause the anchor to coil out and dig into the earth to a depth of thirty feet, effectively anchoring us to the ground. If our calculations are correct, the corkscrew will hold Lucille down, even under the force of an F4 tornado."

"What about flying debris?" Tracey was still nervous. "Don't most tornadoes have a lot of flying debris?"

"They sure do," Sylvia answered. "Many times the color of a tornado funnel can be a telltale sign of how much debris it has in it. A white funnel usually means less debris, and a black funnel can mean lots of debris. The tornado we are chasing right now is still pretty white, so I am hoping that means a low amount of debris. But with her titanium armor, puncture-proof tires, and shark-tank glass—Lucille should be able to handle most debris any tornado would throw at her."

"So, basically what you're saying," Blaine simpered, "is that Lucille is indestructible."

"Something like that," the meteorologist smiled. She punched down the accelerator, causing Lucille to shoot up the turnpike at a tremendous speed. It wasn't going to take them long to pass the F2.

The highway was mostly abandoned, and the vehicles that were there had all pulled off on the shoulders, so the road was wide open for the storm chasers. The rain began to fall even harder. Now, mixed with the rain were ping-pong ball sized hailstones. The white stones pinged, bounced, and ricocheted off Lucille's titanium plating, but left no marks. The tornado remained visible, staying on the north side of the turnpike. They caught up to it, and then passed the whirling menace.

"Come in, Jayman, come in," Thunderstone spoke into the walkie-cell.

"Reading you loud and clear, Sylvia. Go ahead."

"We have gotten ahead of the twister, and are now looking for a suitable place to park and anchor Lucille."

"Sounds good, Sylvia. Go ahead with the plan, but be aware that this may not be the only tornado we see today. The air pressure within this storm remains strong, so it is likely we will see more circulation. I will keep you posted."

"Copy that, Jayman. Over and out."

Thunderstone kept driving straight on the highway for a few hundred more yards. Then she cranked hard left on the steering wheel, pulling Lucille off the turnpike, down an embankment, through a shallow ditch, across a small service road, over a cattle grate, and into a field. Mud was flying everywhere as she drove.

Though the twins were buckled in, they each felt the need to reach out and grab something to hold onto. This was turning into a crazy bumpy ride! Mud, rain, and hailstones were hitting the windshield with such rapidity that the wipers were having a hard time keeping up. Thunderstone continued to guide Lucille through the muddy field, as she kept an eye trained out the side window at the tornado.

Just before they reached the end of the field, Sylvia brought the storm-chasing vehicle to a stop.

"Okay, Sassafrases! I think this is the spot!" she announced. "It looks like we are in the direct path of the tornado!"

Blaine and Tracey were speechless, partly out of excitement, partly out of fear.

Sylvia thought out loud as she began preparations for encountering the tornado. "Exterior sensors on. Floodlights illuminated. Doppler radar transmitting. Corkscrew anchor engaged."

The meteorologist looked at the twins with anticipation in her eyes. "Now all we have to do is wait."

But they didn't have to wait very long because at that moment, the F2 tornado appeared right before them, coming their way like a giant angry blender. The twins noticed the twister was less white than it had been earlier. Now it looked browner. Blaine was about to ask Sylvia if the darkening color was because of debris when his question was suddenly answered by a tree flying out of the tornado through the air, straight toward their windshield. At first it looked like a small branch floating through the air, but as it got closer, it became apparent that it was the entire trunk of a tree.

The tree flew toward them, causing Tracey to scream. Blaine ducked down. Sylvia stared directly at the approaching log. The tree made direct contact with Lucille, but it glanced off to the side of the windshield, hardly leaving a scratch.

"Good job, Lucille," Thunderstone exclaimed, proud of her hardy vehicle.

"And look!" She pointed at a couple of the monitors. "It looks like we are starting to get some data from the exterior sensors!"

The closer the tornado got, the more stuff they could see flying through the air and the more debris hit Lucille.

"What is that pink thing flying in our direction? Is that

a....pig?" Blaine questioned.

Tracey saw it too, but she couldn't believe it. "But...pigs...can't fly."

"It's not really flying. It's being tossed around by the tornado!" Blaine argued. "Wait...is that a pig...wearing overalls?"

"It's Buster!" Sylvia suddenly interjected.

"Buster?" the twins asked.

But instead of explaining, the meteorologist just gazed intently out the window at the rapidly approaching pink flying pig. It got bigger and bigger as it approached, and it soon became apparent that this pig was even larger than the tree that had grazed them earlier. And the pig was indeed wearing overalls. It also had big wide eyes and a large goofy grin.

"What in the world is happening?" Both Sassafrases thought their minds might be playing tricks on them. Had they been sucked up into the tornado and carried off to a make-believe wacky land?

The giant pig, which was as big as a house, slammed into Lucille. And when it did, a shattering sound immediately reached the passengers' ears. Again, Tracey screamed, Blaine ducked, and Sylvia stared directly at the object. But following the shattering collision, there was no getting sprinkled with glass. There was no mud splattering their bodies. There were no strong winds hitting them. They were not being sucked into the tornado. Tracey stopped screaming. Blaine looked up. The shattering sound had been the pig, not the car. Upon impact with Lucille, the pig had fractured into hundreds of pieces.

"In just a few more seconds, we'll be inside the tornado!" Sylvia shouted, but just as she did, the twister took a major shift toward the south and missed them completely. And then, to disappoint the meteorologist, the tornado began to quickly dissipate.

"These twisters can be so unpredictable," the storm chaser sighed, more than a little downcast.

"Poor Buster," Sylvia said, as she looked at the twins.

"Who's Buster?" the twins asked in unison. They were both more than a little curious.

"Buster is the happy-go-lucky mascot of Buster's Barnyard, Oklahoma's largest family friendly home-style restaurant and fun zone."

Thunderstone lifted her hand and pointed across the street. The twins looked, and just over the barbed wire fence and across the road was a Buster's Barnyard. It was a barn-shaped restaurant the size of a giant grocery store. The twins couldn't believe they hadn't seen it earlier. They must have just been too caught up in the tornado and blinded by the rain. The restaurant had bright lights and well-designed signs, but curiously missing from an iron perch on its roof was its huge mascot.

"The Buster that hit us was made out of fiberglass," Sylvia went on. "The tornado ripped it right off the Barnyard's roof and tossed it at us."

The twins were amazed at the power tornadoes possessed. The three sat silently for a few moments and stared at the large restaurant across the street. There were several cars and a couple of buses in the parking lot, but no one could be seen going in or out. Tracey wondered if the restaurant had a storm shelter. Blaine wondered what kind of food they served and what kind of arcade games they had. Both were glad the tornado had disappeared.

"Sylvia! Sylvia, come in!" Jayman's voice on the walkie-cell broke the silence.

"Reading you loud and clear, Jayman. Go ahead."

"Do you currently have eyes on the tornado?"

"No, Jayman, we don't. And we can't. It's gone. It has completely dissipated."

"No, not the F2, Sylvia, the F5! Our radar is showing that an

F5 has formed right behind the F2, and it is basically taking almost the same path the other tornado took!"

"What? An F5? Jayman, we don't see it yet! We don't have eyes on it."

"You will soon, Sylvia! It is over a mile wide and it is coming your way!"

"Jayman, do you think we should...?"

"Don't even think about it, Sylvia. I know what you were going to ask, and you shouldn't do it. Don't even think about it!"

"But, Jayman, if we stay anchored here, we would have such a good chance of getting Lucille and all her sensors inside that twister!"

"I know, Sylvia, and I want all the data as much as you do, but you just can't do it. It's too risky. Lucille is only tested to withstand F4 winds, and according to my calculations, even that rating is questionable. It's just not safe. You have to get out of there!"

"But, Jayman, I—"

"Get out of there, Sylvia!"

"Okay, you're right. You're right, Jayman. We'll start moving now."

"Copy that, Sylvia. Keep me posted."

"Will do, Jayman. Over and out."

"Over and out."

Just as soon as Thunderstone finished her call with her colleague, the meteorologist and the twins caught a glimpse of the huge, wide, menacing F5 tornado on the horizon. It was spinning and barreling toward them like a giant grizzly bear made up of wind, debris, and all Mother Nature's fury. The sight was terrifying. Sylvia quickly started to turn off the wind sensors and disengage the corkscrew anchor, when she was interrupted by a shout from Tracey.

"Look, there are people inside of Buster's Barnyard!"

## Chapter 4: The Congolese Jungle Treasure Hunt

### *Carver's Rain Takes the Lead*

Sylvia and Blaine joined Tracey's gaze across the street. They could see several people emerging from inside the restaurant, frantically exiting the front doors. The people looked as if they were Buster's employees, as they were wearing pink shirts and blue overalls similar to Buster. Many of them were screaming. Some were pointing at the approaching F5 as they all fled to their cars to peel away at breakneck speeds.

"Good, good," Thunderstone nodded. "All the employees must have heard the tornado sirens. They are all getting out of here safely."

"No, not good," Blaine blurted. "Look!"

The Sassafras boy pointed back at the restaurant where, behind one of the front doors with her hands up on the glass and a scared look on her face, was a small girl. Thunderstone looked at the little girl—then at the fleeing employees—then at the idle school buses left in the parking lot.

"They left the kids," she gasped. "They left the kids all alone inside the restaurant."

The meteorologist looked at the roaring F5. "If an F2 can rip Buster off the roof, an F5 will be able to blow the whole restaurant to the ground . . . somebody has to go in and rescue those kids!"

By now, the twister was dark, ominous, and dangerously close. It looked like it was alive! Its strong battering winds had picked up. Hailstones the size of a grapefruit had begun to fall. Lightning bolts crashed. Shaking thunder boomed. The tornado was rapidly converging on the large restaurant.

Suddenly, to the twins' surprise and fear, the meteorologist opened up Lucille's side door and stepped out into the weather. She closed the door behind her. Then, through the window, she mouthed the words, "Everything is going to be fine," to the twins.

Sylvia Thunderstone then set her face like stone, turned toward the storm, and ran directly into it.

In the next second, all Blaine and Tracey could hear was a sound like that of a barreling freight train as the F5 tornado approached. There wasn't time to run. All they could do was wait—or maybe not.

The twins already had all of the data they needed on wind, wind patterns, downbursts, and tornadoes. They didn't have all the pictures they needed, and they hadn't sent all the SCIDAT data to Uncle Cecil yet, but they didn't have to do that anymore in order to progress to the next location because the glitch was gone. They had the freedom to zip away any time they wanted or needed to.

As Blaine and Tracey considered their options, the black winds of the terrible twister wrapped all around them, and Lucille started spinning—not just around and around, but she was going up as well. The powerful tornado was unscrewing the storm-chasing car from its anchored spot in the field!

"Blaine," Tracey screamed. "Should we go? Should we zip to the next location?"

Blaine had an embattled look on his face. "I don't know, Tracey. I don't know! What about Dr. Thunderstone? We can't just leave her behind. And what about the kids she went to rescue? What if they need our help?"

"We aren't going to be around to help if we stay much longer," Tracey argued.

"But we are Sassafrases! And Sassafrases don't quit. We have to stick around. We have to see this adventure through to the end. We have to make sure our friend is okay, don't we?"

Tracey was scared, but she came to the same conclusion her brother did—they should stay.

Tracey nodded her resolve as she reached over and clutched her brother in a fearful hug. Blaine was scared too, so he hugged his sister back. The twins both closed their eyes—was this their last run? Was this the last scientific location Blaine and Tracey would see?

Debris of all shapes and sizes continued to smash and crash against every inch of Lucille's exterior. And then it all stopped.

The Sassafras twins opened their eyes and looked up from their trembling embrace. All they could see outside of Lucille was blue. The F5 tornado was gone. The sky was clear, which was good. But everything else they saw was not.

All the trees in the area were snapped in half, uprooted, or gone. The barbed wire fence that had been right in front of them had completely disappeared, with the exception of a piece that was now wrapped tightly around Lucille's hood. Most disturbing of all was the site where Buster's Barnyard had been—it was now completely gone. The barn-shaped restaurant was leveled. All that was left was the foundation, with piles of assorted unrecognizable rubble scattered across it.

"C'mon, Tracey, let's get out of this car! Let's go help Dr. Thunderstone!"

"But Blaine, just look. How could anybody survive that?"

"I don't know, Tracey, but we have to go look. We have to help if we can!"

Tracey nodded, reached over, and opened Lucille's side door, preparing to step out of the car, but she abruptly stopped. The car was now high above the ground, balancing on its corkscrew anchor. It was teetering and threatening to tip over into the muddy field.

"Blaine, this thing is about to fall over! How are we supposed to get down?"

"How about you jump, little lady," came the answer, although the voice didn't come from Blaine.

It was Old Doc Hibbel! He was in the field below them in his covered wagon, which was being pulled by Ike and Wyatt, his two faithful horses.

"Doc?" the Sassafrases exclaimed. "But how . . . what . . . where did you . . . ?"

"Just jump, you two. Lucille is a-teetering, and you don't have a second to spare."

Blaine and Tracey jumped and landed on the covered wagon's driver's perch next to Sylvester. Just as they did, the corkscrew came loose from the earth, and Lucille toppled over, landing with a thud.

"That was a close one!" Hibbel whistled in relief.

He looked at the downed storm-chasing car for only a second. Then he turned his attention toward the downed restaurant.

"Now, let's go see if we can't find Sylvia and the kids. Heeya!" The old cowboy slapped the reins at the backs of Ike and Wyatt. The two horses immediately pulled the wagon out of the field, across the small street, and up to the pile of rubble. Blaine, Tracey, and Doc Hibbel got down from the wagon and began walking among the piles of destruction. The thought at the forefront of all their minds was, "Where is Sylvia?"

Suddenly, a built-in hatch in the floor opened up and a brazen figure emerged. It was Sylvia Thunderstone and she was followed by an entire class of shivering elementary aged children and a couple of their teachers. She was beaten, bruised, and soaking wet, but she was very much alive! And she had saved every last one of the school children.

The Sassafrases raced forward and each put one of the meteorologist's arms over their shoulders to help her walk.

"I'm so glad you're okay," Tracey said.

"I'm so thankful Buster's Barnyard had a storm shelter," Blaine added.

Sylvester Doc Hibbel looked at the rescued children. Then, he looked at the two teachers who hadn't abandoned them. And then, he looked at the woman who had saved them all.

"You are a true Thunderstone!" he announced proudly to Sylvia. "Just like your great, great grandfather!"

An hour or so later, Blaine and Tracey found themselves zipping ecstatically through swirls of light. They had just survived one of their most dangerous locations to date. When the going got tough, they had pressed on to complete their prairie adventure because they were Sassafrases, and Sassafrases don't quit.

In the end, everyone had escaped being hurt by either of the tornadoes. Lucille had managed to gather new and important data from the F5 with her sensors. And, the twins had gathered and sent in all of their SCIDAT data and pictures. According to the LINLOC applications on their smartphones, they were now riding the invisible zip-lines to the Democratic Republic of the Congo, where the local expert's name was Carver Brighton, and they would be studying the topics of rain, monsoons, thunderstorms, and floods.

Their travel ended with the customary jerk. Their carabiners unclipped from their lines, and they both slumped down to wait for their sight and strength to return.

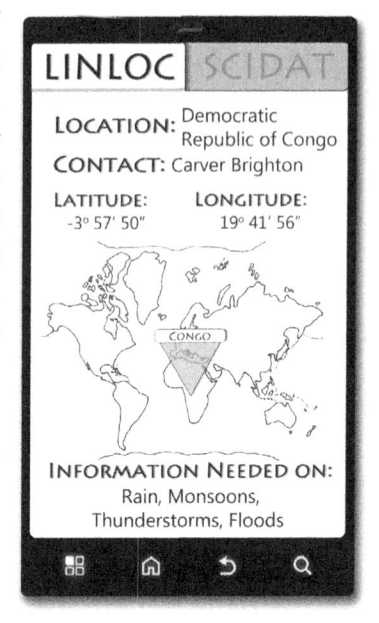

**LINLOC** SCIDAT

**LOCATION:** Democratic Republic of Congo
**CONTACT:** Carver Brighton
**LATITUDE:** -3° 57' 50"
**LONGITUDE:** 19° 41' 56"

**INFORMATION NEEDED ON:**
Rain, Monsoons, Thunderstorms, Floods

Their hearing was the first sense that returned, and when it did, two wretched screams of slightly different tones reached the Sassafrases' ears. When the twins' sight caught up with their hearing, they spied the source of the screaming.

Two men, one black and one white, were staring right at Blaine and Tracey, with wide eyes full of fear. They were both hollering helplessly and hauntingly. The native-looking man dropped his heavily loaded backpack and began to back away from the twins first. The other man, a foreigner who was carrying armfuls of rolled up maps, soon followed. All at once, both men turned and quickly ran off into the surrounding jungle with the maps. Their screams following their path.

Blaine and Tracey looked at each other in confusion. They had never had a landing like this before. Blaine went to move from his landing spot and suddenly realized that he was balancing up high in the air on some kind of stone arch. If he moved too much, he might fall. It wasn't a long, long way down, but it was far enough down to hurt if he landed wrong. The boy took great care to shift his body.

Tracey began to carefully adjust her position as well. What had just happened? Had those two men just seen the Sassafrases appear out of nowhere upon their landing? That wasn't supposed to happen. The lines were designed to land the twins as close to their local expert as possible without being detected. Had one of those men been their local expert? All these questions and more raced through Blaine's and Tracey's minds as they tried to carefully climb down from the stone archway.

The descent proved to be difficult because not only was the arch tangled in vines and covered in moss, but it also had life-size statues of different jungle animals mounted to the top of it. Climbing around and over these stone figures without falling was proving to be a challenge.

Suddenly, the twins' progress was interrupted when they heard two more screams. But this time the screams were shouts with actual words.

"Moving statues!" the two scared voices screamed.

Blaine and Tracey looked and saw four people appearing out of the jungle. Two of them were somewhat plump with very pale skin. They wore khaki clothes with old-time explorer pith helmets. The other two, who were not screaming, were both fit, had darker skin, and were dressed in modern outdoor apparel and wearing baseball caps.

"Moving statues!" the pith-capped individuals screamed again.

"See? I told you!" the shorter of the two said as he pointed toward the Sassafrases. "It's the Njanga Gate, and it is being guarded by moving statues!"

The tan fellow just shook his head and chuckled. "Those

aren't moving statues, Dimsley. Those are kids."

"Kids?" The man named Dimsley screeched like that was preposterous. "It's not kids that are moving, it's sta...wait...It's not statues...It IS kids, just kids. No! This isn't good. You mean to say that our handler and our map expert got scared away by a couple of kids?"

"How can you blame them?" the tan man said, still chuckling. "That fable you told about some ancient gateway being guarded by immortal moving statues had Bemba and Brady scared from the get go. My guess is when they saw the kids climbing around up there on this stone thingy—they saw ghosts, not kids."

"It's not a stone thingy, Brighton! It's a gate. It's THE gate—the one we've been looking for! This is the long lost Njanga Gate! It is the end and the beginning. It is an exit and an entrance."

Brighton, who the twins were supposing was their local expert, rolled his eyes in a good-natured way as if to say, "here we go again" as Dimsley began what sounded like a dramatic and memorized monologue.

"The Njanga Gate that now stands before us is the end of the world as we have known it up to this point! It is an exit from our modern reality and the laws of time, nature, and science that we have become accustomed to! The assortment of stone animal statues that adorn its top serve as immortal sentinels, guarding its threshold with all-seeing eyes. Their eyes watch every step, every movement, and every breath of the few who make it this far. They are ready to pounce if any malice is to be found in the heart of the gate-crosser. Beyond this Njanga Gate is deepest darkest Africa. Very few have ever made it to this gate, and even fewer have stepped beyond it.

"'Njanga' is the Congo word for 'secret path.' This gate serves as the beginning of a world forgotten by time. It is the entrance to the land of the strange, bizarre, and unexplored. It is a place where a rabbit can be as big as a gorilla, and a gorilla can be as big as a dinosaur. It is a place where headhunters hunt, pitfalls await, and mankind is not at the top of the food chain. Faint not, my dear friends, for if you do, you may never rise again. If any courage can be summoned in your soul, summon it now! For you will need every last ounce to step beyond the Njanga gate!"

Upon the conclusion of Dimsley's speech, a light rain began to fall, and Brighton couldn't help but continue to chuckle.

"You have painted quite a mystery for us here at the Njanga gate, Dimsley," he said. "But let me tell you about something a little more set in stone—pun totally intended."

Now Dimsley rolled his eyes.

"Let's talk about this rain that is beginning to fall," Brighton continued, starting a monologue of his own. "Rain, my dear Dimsley, is simply water falling from clouds as droplets. Rain forms when warm, moist air rises and condenses to form a cloud of water vapor. The micro-droplets then collect together to form bigger droplets, which fall to the ground because of gravity."

"Raindrops are quite tiny, only a hundredth to a tenth of an inch in diameter. Very fine drops of rain fall at a rate of about two miles per hour, while very heavy drops fall as fast as eighteen miles per hour. In this area, heavy rain can cause many problems, such as flooding and landslides. In highly populated areas with large cities, there is often acid rain. Acid rain is rain with a low pH due to sulfur dioxide and nitrogen oxides that have been released into the air by factories. Acid rain can damage plant life and even buildings. Out here in the rainforests, however, acid rain is rare.

"Tropical rainforests, like here in the Congo, can get anywhere from eighty to a hundred inches of rain a year. These areas are warm and humid all the time, while a temperate rainforest can have a cool season and sometimes even experience frost. Temperate rainforests usually get around a hundred inches of rain annually. Several different temperate forests also get additional moisture from coastal fog. I bet you didn't know, Dimsley, that the highest rainfall ever recorded in the world was in the country of India, where they had one thousand inches of rainfall in one year."

At the conclusion of Brighton's speech about rain, Dimsley clapped his hands in a sarcastic manner and then quipped, "Oh, wow, Brighton. All your amazingly boring knowledge about rainfall just had me on the edge of my seat!"

"You don't even have a seat, Dimsley."

"I didn't mean a literal seat. I was referring to the coined phrase."

"I know what you meant. I just didn't appreciate your sarcasm."

"Well, you should appreciate it, Brighton, since sarcasm is your native tongue."

### *Moving Monsoons*

"Good day, you two." The other pale skinned man in khaki greeted Blaine and Tracey with a thick British accent. He was dressed like Dimsley, but he was at least half a foot taller. The man had a huge mustache that curled up at the ends, and he was wearing an eyeglass connected by a gold chain to his pith helmet.

"Hello," the twins responded as their sneakers hit the ground, finally down from the stone arch.

"My name is Garfield T. Wellington the Fourth," the man

said, "I am the benefactor of this treasure hunt."

"Treasure hunt?" Blaine asked.

"Yes, my good young man, treasure hunt. We are here in the Congo in search of the Giant Bonobo Diamond."

"The Giant Bonobo Diamond?" Tracey questioned.

"Yes, my darling young lady, the Giant Bonobo Diamond. It is said to be hidden somewhere deep in the jungle in a secluded and dangerous temple. We had a series of maps we were following, but it seems these maps have run off with our map-reading expert, Brady, and our bag handler, Bemba. I recruited five to aid me in this treasure-hunting quest, but it seems that I only have three remaining."

Wellington reached his hand out and put it on the shoulder of the strong, quiet, black man standing next to him.

"This is Bakaza," the British benefactor informed. "He is our Congolese guide and trailblazer. With him, we should be safe against anything that comes our way."

The Brit then pointed at the other two men who were still arguing with each other. "Those two are Carver Brighton and Stuart Dimsley. They have been trying to top each other since the day they met. Both were highly recruited by rival colleges from the same town. Carver has a degree in geochemistry from Marble Bridge Tech, and Stuart has a degree in cultural studies from Driskmon University at Marble Bridge. I hired them for this expedition to be my scientist and cultural expert, respectively. I am very interested in these subjects, and the two of them are doing a good job of helping me to understand topics from their particular fields—that is, if they can just stop arguing.

"It is quite a miracle that we actually found this spectacular Njanga gate. It has been a difficult journey up to this point, even with the help of maps." Wellington paused and looked thoughtfully at the twins. "Since we lost two men, I could really use your assistance as we search on for the hidden temple. I see that Bemba dropped his pack when he fled in fear. If the two of you will assist me, by splitting up the contents of Bemba's bag and carrying them, you will be paid handsomely."

The twins looked at each other quickly, then back at Garfield T. Wellington the Fourth, before answering, "We would love to help you!"

The rich British adventurer shook hands with each of the twelve-year-olds to seal the deal. Once he had, Blaine and Tracey immediately went over and separated the contents of Bemba's bag evenly into their backpacks. As they did, Brighton and Dimsley continued their tussle with words. "Great, Dimsley. I'm so glad you told me all about the tribe that may or may not live in this region. And about their supposed belief that dinosaur species still live in the Congo, but do you know anything about monsoons?"

"I don't want to know anything about monsoons!" Dimsley retorted.

"Well, you should because, unlike your cultural dinosaur myths, monsoons actually exist today! Have you noticed how the rain has picked up? Do you know what time of year it is right now? Do you know anything about the seasonal changes here in the Congo?" Carver peppered his rival with questions.

"What are we playing here, Brighton? Twenty questions? Yes, I've noticed that the rain is getting heavier and heavier! But unlike the falling rain, we can actually be hurt by the dinosaurs that live in the ancient jungle beyond this gate. I am not interested in knowing anything about monsoons!"

"Well, I'm going to tell you about them anyway, Dimsley! Monsoons are the result of seasonal changes in the strong winds affecting a region. The largest monsoon winds occur in Asia, but there are smaller ones along the equator in Africa and the southwestern United States. These winds are the cause of the wet and dry seasons that you find throughout the tropics. The Congo is affected by the Asian Australian monsoon winds, which stretch from the northern coast of Australia up through the Pacific and across the Indian Ocean to the coast of Africa.

"Summer monsoon winds, which happen to be what we're experiencing right now, Dimsley, typically blow from the southwest regions across the ocean and onto land. The air brought onto land is humid, which means it is rich in moisture. This causes heavy, or torrential, rainfall. The abundance of rain during a monsoon season fills wells and aquifers that the people in the region will rely on during the dry season.

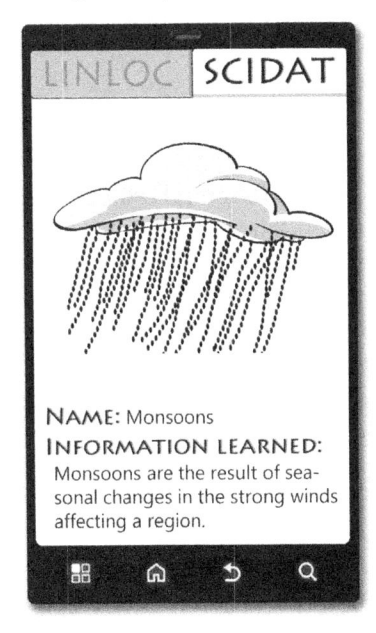

"Winter monsoon winds are opposite from summer ones. These winds typically blow in from the northeast regions, bringing drier air. The exception to this is in the Pacific Northwest, where the winter monsoon winds bring about their wet season. Although monsoon rains are beneficial, they also have loads of potential dangers, like —"

Garfield spoke up before Carver had a chance to finish. "Very interesting, Brighton, very interesting indeed—thank you for sharing. That is what I'm paying you for. And, Dimsley, the bit about the dinosaurs was interesting as well. But if, in fact, dinosaurs still do exist, I want to see them, not just talk about them. So, tally-ho, crew! Rise to your feet! Let's step beyond the Njanga Gate into deepest darkest Africa to see what adventures await!"

Blaine and Tracey stood with full backpacks and followed the adventure troop as they walked through the Njanga Gate. Was it

LINLOC SCIDAT

NAME: Monsoons
INFORMATION LEARNED: Monsoons are the result of seasonal changes in the strong winds affecting a region.

really possible they were now entering a land that had somehow gotten stuck in time? Had it remained unaffected by modern changes that had reached every other part of the world? Were there really headhunters, pitfalls, and giant animals out in the dark jungle ahead of them? The twins didn't know, but they figured if they could survive an F5 tornado, they could survive a mystery stricken jungle during monsoon season. No sooner had the group taken collective steps beyond the Njanga Gate when they suddenly felt themselves falling.

"It's a jungle pitfall!" Stuart Dimsley screamed.

"Not exactly!" Carver corrected. "It's a mudslide . . . caused by . . . the rain!"

Regardless of semantics, the entire group of six were sliding down fast, one after the other. The dirt gave way to mud and then the mud gave way to running water. The twins could tell they were at the top of a sizeable jungle slope. And that there was still a long way down before they reached the bottom.

Blaine and Tracey found themselves actually having fun as they rode down the monsoon-created jungle flume. Some of the others in the group were screaming, but the twins were grinning from ear to ear.

The muddy mountain waterslide was a straight shot down at first, but after a couple of hundred feet, the channel the six were in branched off in several directions. Bakaza took a route to the far right. Wellington took the straightest route Brighton and Dimsley careened off to the left together. Then, Blaine and Tracey went down even further left. The twins hoped they weren't lost as they continued to slide down. The further down they slid, the stronger the force of the rushing water became, and the less control the twins had over the movement of their bodies.

Blaine, who was in front of Tracey, started spinning wildly. Tracey actually started doing occasional accidental somersaults. There were blind curves and violent twists and turns. There was more and more vegetation closing in on the sides of the flume. There were even plants starting to arch completely over the top of the flume, turning this monsoon-caused waterslide into more of a tunnel-slide.

Tracey opened up her mouth at the wrong time and swallowed an entire mouthful of muddy water. Blaine kept getting slapped in the face by roots and thorns that were hanging down. The twins were no longer having fun. They were now just hoping to survive.

Swish, gurgle, splash, boom, slap, slide, gurgle, splash, swish! Down and down the Sassafrases went, until they both felt themselves go airborne for a split second before their final splash.

The wild water ride was over . . . or was it? The twins had landed in a river—a huge river. Each of the bodies belonging to the members of their expedition party made a splash as they exited different flumes and landed in the river close to the Sassafrases.

"Out! Get out of the river!" Stuart began yelling, almost before he even came up out of the water after his landing.

"Why?" Carver inquired. "Why do we need to be in such a hurry to get out of the river?"

"Because of the catfish!" Stuart answered frantically.

"Because of the catfish?" Carver laughed. "Catfish are harmless."

"Not these catfish, Brighton! This is the Congo River, and the catfish here are big enough to swallow a full-grown man, and they're mean and nasty enough to do it. They will attack us as if they are piranhas and swallow us like we're Jonah. The bottom of this river has got to be teaming with these ferocious creatures!"

Carver remained skeptical. Bakaza's face stayed emotionless. However, Blaine, Tracey, and Garfield T. Wellington the Fourth were scared, and their faces showed it. The three of them chopped at the water with strokes as big and fast as possible, desiring very much to get to the river's edge as soon as they could. They followed behind the fleeing Stuart Dimsley. They were all managing to swim very fast, which was good. But, they were all swimming in the wrong direction—out into deeper water—which was not good.

"Dimsley! Where are you going?" Carver called from the spot where he and Bakaza had already reached the shore. "I thought you wanted to get away from the gargantuan catfish, not swim deeper into their feeding grounds!"

"Oh, my," the twins heard Stuart say in shock. "Oh, no! No! No! No! Swim back! Swim back! We've been going the wrong way!"

The four swimmers each did a one-eighty and began clawing through the water again, this time in the correct direction. The river water was muddy and murky, making it impossible to see anything that might be swimming down below. The twins' imaginations ran wild with visions of catfish as big as school buses with wide open mouths swimming up to eat them. They continued to churn for all they were worth toward the shore.

What was that?

Had something bumped against Tracey's leg?

Now Blaine felt it too. Something had definitely made contact with his left shin. Evidently something had hit Garfield T. Wellington the Fourth as well because he was shouting about it.

"Ahhh! Collywobbles! Something just swiped against my arm!" he screamed.

The shore was getting closer and their motivation was running high. Garfield, Blaine, and Tracey reached the river's edge before Dimsley. The five on land now reached with outstretched arms, imploring Stuart to finish his swim. He was still about five yards out and was making good progress when suddenly he let out an anguished shout of pain.

"Ahhh! It's got me! Oh, the horror!" Stuart was screaming and crying out in pain. None of the five on the shore could see what had gotten him.

"The agony! The pain!" He continued to scream, as he flailed and flopped and finally made it to shore.

The five collectively pulled him up out of the water, and it soon became apparent that he had indeed been attacked by a fish. But it was not a catfish. Somewhere, somehow, Stuart had lost his left sock and shoe, and now, clamped to his big toe, was a tiny fish. It was actually surprising that this fish's mouth was even big enough to fit around Stuart's toe.

The cultural expert continued to cry out in pain until he looked down at his foot and saw his "attacker."

"Well, isn't that cute?" Carver chuckled. "It's a little perch, but I guess they are pretty ferocious, aren't they, Dimsley? Even more ferocious than the ancient giant catfish."

Stuart pulled the little fish off his big toe and tossed it back into the river. He looked up at his rival, Brighton. He was about to attempt a witty comeback, but instead his face froze in horror. It was immediately apparent that he wasn't staring at Carver but at something behind Carver.

Brighton, Bakaza, Wellington, Blaine, and Tracey slowly turned from Dimsley and the river to look into the jungle behind them. And what they saw painted looks of horror on their faces as well, with the exception of Bakaza, whose expression never seemed to change. There at the jungle's edge, with spears and machetes in their hands, was an entire tribe of pygmy headhunting warriors.

## Chapter 5: The Search for the Giant Bonobo Diamond

### *Thundering Pygmy Warriors*

Everyone remained frozen for what seemed like ages.

Not one of the pygmy warriors twitched or moved a muscle. The six treasure hunters were barely breathing, much less moving.

Maybe Stuart Dimsley had been wrong about moving statues and giant catfish, but it looked like he had definitely been right about the headhunters.

Unlike the legendary giant animals of the deep Congo, these warriors were very small. The adult males that surrounded them only stood about three or four feet tall. The warriors were at least a foot shorter than Blaine and Tracey, who were both already nearly five feet tall at twelve years of age. However, the small stature of these warriors did nothing at all to make them look less fierce. There had to be at least thirty of them, and every last one of them looked like they could take down a carnivorous dinosaur with their bare hands. They were each wearing animal skin loincloths and colorful beaded necklaces. But more than that, they were all showcasing weapons!

"How long was this stare-down going to last," the twins' fear-stricken minds asked. "How long before the headhunters made their move? How badly was it going to hurt? Was this really how they were going to meet their end?"

Suddenly, a move was made. The shortest among the pygmy warriors let his face change from stone cold frown to bright happy smile. "Hello, my friends," he said cheerfully. "I see you have found our fishing hole."

"Your fishing hole?" Stuart questioned. "You mean, your weapons aren't for . . . our heads?"

"No, of course not!" the shortest pygmy laughed. "We use our spears and machetes for fishing."

"Oh," eeked out Stuart plainly, as his five companions looked at him with raised eyebrows of growing skepticism.

"My name is Chief Wazabanga. And this is my tribe." He motioned to his companions. "Today, we will not fish. Instead, we will take the six of you back to our camp and we will share a meal."

"That sounds excellent!" Garfield T. Wellington the Fourth spoke up. "We would be honored if you would have us for a meal."

The six treasure hunters soon found themselves marching through the jungle with about half of the pygmies in front of them and half behind them. After a short hike, they reached camp. Upon arrival, they were each seated on stumps close to a crackling fire. Wazabanga stayed by the fire to talk with his guests, while the rest of the men disappeared off somewhere to prepare the meal.

"What brings you to our jungle?" the pygmy chief asked.

"We come in search of the Giant Bonobo Diamond,"

Wellington answered. "We lost our map expert earlier on our journey, but we feel as though we are still headed in the right direction."

"Oh, yes," Wazabanga replied, nodding his head. "The Giant Bonobo Diamond. I have heard the legends surrounding this prized jewel. If the secret temple where they say this diamond is hidden truly exists, I suppose you are going in the right direction. What are you going to do with the diamond if you find it?"

Wellington paused before he answered, then responded," I guess I really haven't thought that far in advance. I have been so focused on the hunt."

Chief Wazabanga smiled. "Ah, yes, the hunt. That is what thrills us all the most."

The small man looked off into the jungle toward the spot where most of his tribesmen had disappeared. "And when we hunt, it is better for us to have full stomachs, is it not?"

The chief made a signal with his eyes toward his men. "And speaking of full stomachs, here comes my tribe now."

The pygmy warriors rushed out of the jungle, each with a machete in hand.

A sudden fear made Blaine's and Tracey's stomachs jump.

What was going on here? Were they being attacked by their hosts?

Before the group of treasure hunters really even knew what was happening, they found themselves being . . . served.

Each of the pygmy warriors were holding their machetes sideways. And on their weapons were a sizeable servings of what looked like fried rice.

"We call this MVM for 'Machete Vegetable Medley,'" Chief Wazabanga announced proudly. "It is a delightful array of sweet and spicy vegetables mixed with gluten free rice and served on the side of a machete. Eat and enjoy!"

Blaine and Tracey each took a horizontal food-filled machete and enjoyed their MVM along with everyone else in the group. They also enjoyed the good-natured company of the hospitable pygmy tribe as they ate their meal.

The rain, which had been constant since the Njanga Gate, lulled a little during the meal with the pygmy tribe. However, as soon as everyone was finished eating, it started raining rather hard, and this time the rain was accompanied by streaks of lightning and claps of thunder.

"The Democratic Republic of Congo has the highest frequency of thunderstorms in the world," Carver Brighton informed from his stump in between Garfield T. Wellington the Fourth and Chief Wazabanga.

"Tell us more," Wellington bellowed in his British accent.

"Thunderstorms can occur at any time," the Marble Bridge Tech alumnus continued. "Just so long as there is moisture, unstable air, and a lift in the atmosphere. However, they are most likely to occur during the spring and summer months in the afternoons and evenings. This is especially true in tropical rainforests. Tropical rainforests exist near the equator, which is an imaginary line drawn around the earth equally distant from both the North and South Pole. It divides the earth into northern and southern hemispheres."

As Carver spoke, Wazabanga nodded his head with the same amount of joyful interest as Wellington did.

Brighton continued. "Thunderstorms are storms with lightning and thunder. These storms can produce hail and heavy rain, but the biggest danger from a thunderstorm is the lightning. Lightning, which is an electrical current, is produced by the thunderstorm. It is basically electricity in the form of a bright flash across the sky. Lightning is produced as the ice crystals within a thundercloud rub together to produce electrical charges. The negative charges in the cloud are attracted to the positive charges found on the ground or in the air. When the two connect, a lightning bolt forms.

"Thunder is the result of a lightning bolt striking, which causes a small pocket of air to open up. When the bolt is gone, the pocket collapses and creates a sound wave that we can hear. Even though the two events happen close together, we see lightning before we hear thunder because light travels faster than sound."

**NAME:** Thunderstorm
**INFORMATION LEARNED:** Thunderstorms are storms with lightning and thunder.

Blaine and Tracey used their phones to snap pictures for SCIDAT of both the thunderstorm and the rain. Then, they found a picture in the Archive app that worked for monsoons. As usual, they both took precise mental notes while their expert spoke so they could correctly enter all the data into the SCIDAT app at a later time. The Sassafras twins were becoming masters at retaining and regurgitating information. Tracey also noticed her brother's brain seemed to be back to normal after he had acted so strangely back in Oklahoma City at the beginning of the earth science leg of their journey.

The six treasure hunters spent an hour or so with the hospitable pygmy tribe before continuing their journey off into the soaked and soggy jungle in search of the Giant Bonobo Diamond. As they trudged on, the one-shoed Stuart Dimsley talked almost non-stop. The Sassafrases were still having difficulty deciphering what was fact and what was legend.

## Chapter 5: The Search for the Great Bonobo Diamond

"The Congo River Basin covers nearly three hundred and ninety thousand square miles," the Driskmon University graduate was currently saying. "The river cuts right through the two countries from which it derives its name: The Republic of the Congo and the Democratic Republic of Congo. Tens of millions of people live in these two countries, but by far the hugest percentage of them live in the large cities in the south, leaving the vast areas of tropical jungle in the north virtually uninhabited. The lush jungle of the Congo has been left untouched for thousands of years. This mysterious place with all its wildlife is just as it has always been. This includes the hidden temple we are seeking. It's the last place on earth to remain unaffected by time!"

As Dimsley continued to talk, Bakaza continued to hack. The quiet Congo native led the group, constantly swinging a machete, chopping jungle growth out of the way, and creating a path so the group could progress. Bakaza made the task look effortless, but the twins were sure no one else in the group could do what Bakaza was doing, even if they had a chainsaw.

"Surely, we are the first humans to set foot in this place for thousands of years!" Stuart continued. "Or possibly even ever. We very well could be taking steps right now that no man has ever trod! Be alert! Be on guard! For, in essence, we are no longer in our world. We have entered an ancient and forgotten world ruled by different natural laws. A timeless world ruled by giant beasts like dinosaurs!"

As if on cue, just as Dimsley said the word "dinosaurs," a huge crash was heard off in the jungle to the group's right. The sound was loud, but then it stopped as quickly as it had began.

Everyone took a deep breath—maybe it was nothing.

But then the crashing started again, and this time it didn't stop. It definitely sounded like an animal—a large animal—a large charging animal. Could Dimsley actually be right?

Bakaza continued to hack a path as the unknown beast

continued to charge closer. Suddenly, a huge spiky horn exploded out of the jungle just behind where the explorers were.

"Dinosaur!" Stuart screamed. The man then attempted to race up to the front of the line.

"Rhino, not dino," Carver corrected.

"What?" Wellington asked.

"It's not a dinosaur. It's a rhinoceros!"

Bakaza picked up the speed of his trail-blazing, while the rest of the group bunched up behind him. Blaine and Tracey glanced back and saw that the rhinoceros was now stomping and huffing around on a trail they had just forged. The group of six anxiously pushed forward at the speed of a swinging machete as the one-horned rhino continued to crash around behind them. The group couldn't tell if the animal was chasing them or if it was just holding its ground. But whatever was happening back there, they were going to push forward.

All at once, Bakaza hit an open spot in the denseness, and because he was going so quickly and was being pushed from behind by his five companions, when he hit the spot, he stumbled forward. The group then fell in behind him and found themselves sliding down a jungle slope, although this time there was no waterslide. It was just a rough and bumpy tumble through mud, sludge, and an array of plant life. Down, down they rolled down the hill, until they landed in a heap at the bottom of a shallow valley.

Everyone immediately looked up to see if the rhinoceros was following. After quite a long stare and an extended listen, it became apparent the beast was not after them anymore. Everyone breathed easily as they stood up to wipe mud off and check on bruises.

"It really exists!" Stuart exclaimed. "The lost temple of the Giant Bonobo Diamond!"

The twins expected Carver to have a rebuttal or correction,

but he remained silent. Even Bakaza wore a look of surprise!

"Well done, gentlemen," Wellington said proudly. "Well done indeed! Bob's your uncle, we've actually found it!"

### *Flooded Findings*

The structure the six were now gazing at was a little farther down from where they were. It seemed to be built into the valley itself. It was fairly large and made out of dark gray stone. There were vines and other growth wrapped all over it, so it was mostly camouflaged. But it was definitely there. It was definitely real—maybe Stuart had actually been right about something.

"Why are we all just standing around, chaps?" Garfield asked. "Let's go have a look-see."

"Wait, wait, wait, Sir, wait." Dimsley held up both hands with his palms out. "We must approach with caution. It is quite possible that his temple is guarded by all sorts of snares, pitfalls, and booby traps!"

This statement made Carver roll his eyes. "Booby traps? Really? The bigger problem we are facing right now is the rain."

"Oh, c'mon, Brighton!" Dimsley said in exasperation. "Booby traps are a lot more dangerous than a little rain."

"That's just it, Dimsley. It hasn't been just a little rain; it's been a lot of rain. It has been raining on us virtually the entire expedition. Just look. Look at the entrance to the temple. There is already a couple feet of standing water there, and it's still raining. The temple is built at the bottom of a valley, which makes it in jeopardy of filling up with water. What if we all go into the temple and get trapped inside because of a flash flood?"

"A flash flood? That's ridiculous," Dimsley said, reciprocating his rival's eye roll.

"Not only is it NOT ridiculous, Dimsley, it is quite possible!"

**NAME:** Floods
**INFORMATION LEARNED:** Floods are the result of heavy rains or lots of melting snow.

Brighton exclaimed. "Floods are the result of heavy rains or lots of melting snow. When this happens, the rivers and lakes rise above their normal levels and the water spills out and over onto the surrounding land. Floodwaters move slowly across flat lands, but speed up through canyons and valleys. And those floodwaters can be very dangerous.

"A little more than six inches of water can knock people off their feet and potentially sweep them away. For your information, Dimsley, floods are one of the top weather-related killers. Some floods can be predicted several days in advance, but flash floods come up with little to no warning. A flash flood can occur after a period of intense rain from a slow moving thunderstorm, very similar to the conditions we are experiencing right now.

"The water within the rivers and lakes rises rapidly and without warning and you have a flash flood." Brighton paused for a moment and gestured around at the group's surroundings. "We are close to a river. We are in a valley. We have experienced lots of rain. The conditions are perfect for a flash flood, or at least for some form of flooding. It's my expert opinion that we exercise caution if we choose to enter this temple. Simply because of the flooding factor, not because of any supposed booby traps."

Dimsley was about to counter-attack, but was interrupted.

"Thank you for the information and concern you just expressed, Brighton. Thank you for yours as well, Dimsley," acknowledged Garfield Wellington the Forth, with a tip of the brim of his pith helmet.

"We stand now, gentlemen, in a place we will never stand again," he said with a stern face, as he stroked his large mustache. "Yes, there are dangers ahead of us. Yes, there are obstacles and unknowns, but we have made it this far, which is quite possibly farther than anyone has come before. So, we will not stand here debating. We will not shrink back in fear. We will press forward and grab adventure by the horn. We will take whatever the deepest darkest jungles Africa throw at us. We will enter this hidden temple, come pitfalls or high water, and we will seize the treasure that is within. We will hold the Giant Bonobo Diamond in our hands!"

With that, the big pale man turned and walked straight down toward the entrance of the temple. The remaining five all looked at each other as if collectively asking, "Should we follow him?"

The answer came to each without a word being spoken. Soon Wellington, Bakaza, Brighton, Dimsley, Blaine, and Tracey all stood at the temple's entrance, staring into the darkness inside. The entrance immediately led into a tunnel, which sloped slightly downward. As Brighton had noticed, there was standing water, in which each of the explorers now found themselves ankle deep. The stone on the exterior of the temple was covered with impressive carvings of bonobo monkeys, but what all six were looking at right now was something much more curious. It was located a few yards inside the temple, on the left wall, just at the edge of the available light.

"What is that?" Carver asked. "Is that a spear tip?"

The twins thought it might be an arrow, but whatever it was, it was sharp and it was protruding out of a hole in the stone wall, like it was aimed and cocked and ready to shoot anything or anyone that tried to pass it.

Dimsley looked around proudly at everyone, as if to say, "See, I told you there were booby traps."

Wellington was the first to take an actual step into the temple. Evidently, he wanted a closer look at the protrusion.

"Wait, sir, wait!" Stuart stopped the old Brit. "Let me check it out first."

Dimsley walked slowly through the water up to the curious spike. The other five followed right behind him. The farther they stepped in, the more their eyes adjusted to the dim light. It soon became apparent there was not just one tip sticking out of the wall, but dozens—all set at varying heights.

"What kind of sadistic place is this?" Carver gasped.

Stuart examined the sharp object without stepping in front of it.

"It is definitely a piece all its own," he said, "meaning it's not connected to the tunnel wall . . . It's about two inches in diameter . . . And its tip looks to be covered with some sort of black resin."

The cultural expert let his eyes wander from the one spike to look around at everything else.

"I don't see a trip wire anywhere . . . or any other kind of visible triggering mechanism . . . Man, there's got to be fifty or sixty of these things . . . What are they? Maybe . . . oh, wait! Look! There is some script written at the top of the wall!"

"Can you read it?" Wellington asked.

Dimsley studied the script for a few moments, and then made an attempt at translation. "All who enter here . . . must take a . . . sacred pig?"

"Sacred pig?" Carver questioned.

"No, no, wait . . . a sacred peg. All who enter here must take a sacred peg," Stuart corrected himself. "These protruding spikes must be sacred pegs."

"How can you be sure?" Wellington questioned. "Does the script say anything else?"

THE SASSAFRAS SCIENCE ADVENTURES

Dimsley looked up and studied the strange words a little more and then read:

*All who enter here must take a sacred peg*
*They will be your light*
*When the water is right*
*They will extend your wall*
*When you face the fall*
*They will be your wheel*
*When you seek the jewel*
*Heed the voice of the ancient bonobo.*

The group stood in awed silence for several long seconds in now shin-deep water. If Stuart had translated correctly, they needed to each pull one of the pegs out of the wall. If Stuart had translated incorrectly, who knew what could happen.

"By the looks of it," Stuart said, "there are no empty holes in the wall, meaning that none of the pegs have ever been pulled out. We very well could be the first people to ever set foot in this place, which means that if there ever really was a giant diamond in this temple, it's probably still here!"

The cultural expert then looked directly into the eyes of the benefactor. "Mr. Wellington, shall we proceed?"

Garfield answered by stepping over to the wall and grasping one of the stone pegs with both hands.

"Collywobbles!" the British man shouted as he pulled the peg out of the wall with great force. Carver Brighton and the twins all started to duck, expecting something bad to happen, but nothing did.

"Well, I'll be a dapper dog," Wellington laughed, holding the foot and a half long stone peg victoriously in his hands. "Dimsley, you must have translated correctly! Everyone grab a peg and let's get going down this dark waterlogged tunnel!"

The five followed Garfield's lead, each grabbing pegs, and

then turning to wade deeper into the tunnel. Because of the darkness, it was impossible to see how long the tunnel actually was. The water was now up to the twins' knees and only getting deeper.

Suddenly, several large square holes opened up near the tops of both walls in between where they were and the tunnel entrance! Out of the holes came torrents of water, pouring into the tunnel and adding to the large amount of the water already there. Before any of the six explorers could fight their way out, a strong current was created that pushed them deeper into the temple. Then, to make the situation even scarier, the entrance to the tunnel slammed closed, entrenching the six in strong rushing water and complete darkness.

"What happened?" Wellington shouted. "Is this because we pulled out the pegs? I thought we were supposed to remove them! Is this one of those flash floods? What is going on? The script said that everyone was supposed to take a peg, right? What did it say next?"

"'They will be your light, when the water is right,'" Tracey shouted, answering immediately.

"But what does that mean?" Wellington asked.

Nobody answered.

The powerful current continued to push, and the water continued to get deeper and deeper. It was now up past the twins' waists . . . then their sternums . . . then their necks. Blaine and Tracey now felt their feet leave the floor and they had to swim just to keep their heads above water. The water continued to push and flow.

"Ouch!" Blaine yelped. His head had just hit the tunnel's stone ceiling.

Grunts began coming from the others in the group as well, as their heads hit the ceiling. This wasn't good. They were running out of room to breathe, and the tunnel was continuing to fill up at an extremely fast pace. Blaine was on the verge of panic.

There was now only about three inches of air space left to breathe in. Blaine leaned back and placed his face flat against the ceiling and began sucking in every breath as though it was his last.

The water began filling in around his face, and then it covered it. Blaine was now completely underwater. This is how it was going to end, he thought. Blaine wondered if Tracey was as scared as he was. The strong current continued to push his body along quickly down the black tunnel. Blaine wondered how long he could hold his breath.

Suddenly, there was a shift in direction—the current was pushing up. All at once, Blaine's head burst up out of the water. He gratefully sucked in some air, reached out, and felt a wall—a rough wall that felt like it was made out of sandpaper. He realized that the water was indeed pushing him up, straight up, at a rapid pace.

Blaine heard someone gurgle and cough on the other side of him. Then he heard Garfield T. Wellington the Fourth's voice ask, "Is everyone here? Is everyone ok?"

Before anyone answered, there was a sudden spark of light. It sparked and then went black again. Then, it sparked a second time, even brighter, and this time the light remained. Near the spark was Tracey's face. Blaine was relieved to see that his sister was fine!

"Everyone rub the pointed black ends of your pegs against the walls," the Sassafras girl directed. "The black resin is flammable!"

Blaine could see that they had all made it through the water-filled tunnel and into the shaft. As the water pushed them rapidly upward, they placed the resin covered ends of their pegs hard against the sandpaper-like walls, and sparks began flying from each peg. The sparks quickly gave way to full-fledged flames, almost like they had just lit large stone matches.

"'They will be your light, when the water is right,'" Tracey quoted the script again, sheepishly.

The orange glowing light was a welcome sight after being drenched in darkness. The water continued to push the six up, causing them to feel like they were in some sort of weird swimming pool elevator. The vertical shaft they were in was about the same width as the first tunnel had been, and now, with the presence of light, they could also see that there was an end to this tunnel—a dead end, and they were approaching it quickly—too quickly.

"The rising water is going to smash us against the ceiling!" Garfield cried.

They were already only twenty-five feet away from the ceiling . . . now twenty feet . . . ten feet . . . everyone braced for impact and took in deep breaths, thinking they would soon be underwater again, but none of that was needed because before they hit the ceiling, the direction of the water shifted to horizontal again, taking them into a tunnel they hadn't seen.

All their flames were still lit. All their heads were above water, which was good, but now it was like they were riding on the front of a tidal wave. The water kept moving fast, as they sped forward down this new tunnel.

"What is that?" Wellington pointed ahead. "Is that a pit or a drop-off of some kind?"

Everyone looked and saw that the benefactor was right—the floor ahead gave way to a big black hole. They watched as some of the rushing water ahead of them spilled down into the hole. Blaine assumed that the same thing was about to happen to them if they couldn't find a way to stop. He thought it may be possible to get his feet down to the floor of this new tunnel, if only the wave they were riding would get shallow enough in time.

The force of the water seemed to be weakening a little, but was it happening fast enough? Blaine managed to get his feet connected to the ground, and then he dug his heels in, trying to put on the brakes. It was no use. The water was still too strong. There would be no stopping this ride. The black pit was now dangerously close. Were they all about to shoot over the edge into the darkness?

"'They will extend your wall; when you face the fall,'" Tracey suddenly shouted. "Look! There are small holes in the walls! Everyone jam your pegs into a hole and hold on!"

A rumble of understanding echoed from all the men in the group as they each comprehended the Sassafras girl's instructions.

It was easier said than done. The stone pegs were heavy. The holes they were trying to jab the pegs into were small. And they were riding a wave of rushing water while attempting to accomplish the feat.

Blaine lifted his peg and watched as holes in the wall zipped by. He figured he was going to have to time this perfectly. The boy singled in on a hole up ahead and then waited…waited…and jabbed!

Clink, thud.

Blaine had missed the hole by at least an inch. He focused on another hole; he aimed, stabbed, and missed again! Blaine didn't see any of his companions in the moving water next to him. He hoped that meant they had all gotten their pegs jammed into the wall. The pit was just feet away. Blaine knew he had only one more try at this thing.

Blaine shouted out and swung his arm in a wide frantic arch toward the wall. The water pushed him out over the dark pit.

Swook! His stone peg sunk into a hole he hadn't even been aiming for! Gravity grasped at him, but he didn't let it win.

Blaine hung onto his peg, with one end of it stuck securely in the wall and the other end still ablaze. He craned his head back and saw five more torches. Everyone had managed to 'extend the wall.'

The flow of the water continually got weaker and weaker, and after just a couple of minutes it turned into more of a trickling stream than a raging river. The five that were further back in the tunnel pulled their flaming pegs out of the wall and hurried toward the pit to see if Blaine was okay. He was fine, but he was getting extremely tired of hanging there. He had already tried to reach his foot back in an effort to reach the edge of the tunnel floor, but it was a little bit too far.

"What are we going to do?" Wellington cried out in concern. "How will we rescue this poor lad?"

Without saying a word, Bakaza leaned out over the pit and jammed his peg into a hole that was halfway between the edge and where Blaine was hanging. The strong Congolese man pushed himself safely back. Then, he gestured for Blaine to use this new peg to come back to the edge.

The Sassafras boy eagerly followed Bakaza's suggestion. He swung from his peg to Bakaza's and then safely back to solid ground. After a round of pats on the back from everyone in the group, Blaine and the other five all stared at the large pit before them. The tunnel floor continued on the other side of it, but the pit had to be at least eighteen feet across, and no one wanted to even think about how deep it might be.

Wellington broke the silence by addressing Tracey. "Young lady, your brilliance has shone brightly today. It was you who figured out how to light the torches and how to save us from falling. Do you now have any ideas on how to get safely over this pit?"

"Monkey bars," Tracey responded confidently.

"Monkey bars?" Garfield questioned.

"Yes, sir," Tracey confirmed. "The holes in the wall extend out over the pit. If we can jab our pegs into the wall across the span of the pit, I think we can make it all the way across by swinging on them like monkey bars. Bakaza and Blaine have already placed the first two pegs for us. The pegs are heavy and a little difficult to place, but I think we can do it!"

"Young lady, I think you are right!" Wellington responded.

The British man then turned toward Bakaza to ask him if he would be willing to swing out and place all the pegs, but in silent initiative, Bakaza had already started to reach for the remaining four stone dowels.

The five watched as the strong Congo native fearlessly swung out over the deep dark pit and place the pegs. He safely reached the other side, and then he beckoned for his companions to follow him.

"Ladies first." Wellington gestured toward Tracey.

The Sassafras girl gulped but didn't hesitate. She leaned out and grasped the first peg, took a long swing to the next peg, and then the next. She bravely reached the third peg, just getting the tips of her fingers on it. Tracey felt herself slipping but somehow managed to get her other hand on the peg before she fell. Bakaza had made swinging over these flaming pegs look easy, but they had to be at least three feet apart. Plus, he was much stronger and had a much wider wingspan than anybody else in the group, especially Tracey.

The Sassafras girl gulped again, regained her composure, swung, and reached for the fourth peg. Tracey made quick work of the fifth and sixth pegs before jumping to the other side of the pit. Bakaza was there waiting to make sure she didn't slip.

Blaine followed his sister, without even the slightest mishandling of the pegs. Wellington followed Blaine, and because of his plumpness, he struggled a bit, but with adrenaline fueling him, he made it. Dimsley was next, and he crossed the flaming monkey bars in the same fashion as the benefactor had. Carver Brighton crossed last, showing his athletic ability by retrieving the pegs on his way across. One by one, he pulled the pegs out of the wall and tossed them to his companions. The local expert's boots landed with a solid thud as he swung from the last peg safely to solid ground.

"Dimsley, I thought you were crazy," Carver said, as he reached back and pulled out the last peg. "I wasn't believing or giving any weight to the legends you shared. And while some haven't exactly panned out, this temple is real. And the script you read at the entrance was accurate. This is all pretty amazing."

Carver then gestured toward Tracey. "Luckily, this girl here has figured out what every line of the script has meant so far. I'm sure she will continue to do so."

Tracey smiled, flattered by the local expert's words.

"'They will be your light; when the water is right,'" Brighton recalled the script. "'They will extend your wall; when you face the fall . . . what was the next part?"

"'They will be your wheel; when you seek the jewel,'" Dimsley answered.

"That's right!" Brighton agreed. "I don't know what that means, but I bet we will figure it out!"

The twins could tell that Stuart was appreciative of Carver's kind words.

"We will figure it out, Brighton," Dimsley said, encouraged. "We will indeed. I, too, thought you were crazy with all the science in the past. But, really, I have appreciated the information you have given us about the weather here in the Congo. I will admit, if it wasn't for the floodwaters you talked about, we never would have been propelled through the temple."

Brighton and Dimsley gave each other respectful nods. Carver then took his torch, walked to the front of the group, and kept going a few steps further down the tunnel.

"It looks like the passageway ahead takes a turn to the right," he said. "Dimsley, at this point I believe you. And if you're right, like I think you are, there is a Giant Bonobo Diamond waiting for us right around this corner."

## Chapter 6: Parachuting into Patagonia

### *Snowy Set Downs*

"There is no jewel here! There is no Giant Bonobo Diamond! There is nothing! It is a dead end!" Garfield T. Wellington the Forth was visibly flustered.

Tracey Sassafras, however, was not. She had figured out what the first two lines of the ancient script meant, and she was bound and determined to figure out the third as well. "They will be your wheel when you seek the jewel," she repeated to herself in her head.

The girl looked around at all the stone surfaces, the walls, the ceiling, and the floor. She didn't see anything but flat stone, except, maybe the outline of a large circle on the floor? Tracey kneeled down for a closer look. Sure enough, there was the ever so slight indention of about a four foot circle on the floor. How in the world were they supposed to use their stone pegs with this? She was about to inform the group of her find, when Blaine suddenly shouted out.

"Hey, look! There's a big circle on the wall!"

Everyone ran over and huddled around Blaine who was standing against the dead end wall. The boy had found the outline of a circle there that was virtually identical to the one that Tracey had found on the floor. So now they had two circular outlines, but what were they supposed to do next? How could the stone pegs become 'their wheel?'

Carver Brighton took his hand and began wiping away dust and cobwebs from the circle on the wall. When he did, it revealed one small hole inside the circle at the top and center. Everyone gasped, knowing the small hole was the perfect size to stick one of the pegs into.

Wasting no time, Carver placed his lit stone peg in the solitary hole. He then stood parallel to the wall and placed both hands on the peg and pulled with all his might. Nothing happened.

"Maybe push?" Wellington suggested and questioned at the same time.

Carver reversed directions and pushed forcefully. Almost immediately, with a grind and a creak, the whole four-foot circle began to turn. To everyone's amazement, when the circle on the wall began to turn, the circle on the floor began to rise.

The Sassafrases, the khaki-wearers, and the Congo native watched in awe as Brighton kept churning the circle. As his muscles worked, the peg revolved, and the four-foot circle turned. It almost looked like he was cranking a huge wheel.

"'It will be your wheel when you seek the jewel,'" Tracey exclaimed.

Everyone was staring at the circle that was coming up from the floor. What was underneath that circle? Could it be the priceless jewel they had been seeking? Or were they facing another trap? They would soon find out.

It became apparent that the circle was a sort of stone lid. As it rose they could see that there was open space now being revealed beneath it. The lid continued to rise, supported by four stone braces, and the open space below soon became illuminated with dazzling light. Carver cranked on the wheel faster now, excited by the sparkling light. The stone structure continued to rise until it was sticking up about five feet out of the ground.

Creak. Thud. The wheel on the wall came to an abrupt stop. Brighton could push no more, but it didn't matter. He didn't need to. Because it was right there in this ancient stone display case . . . a huge sparkling jewel—the Giant Bonobo Diamond.

A few hours later, the Sassafras twins were seeing bright, sparkling light, but it was no longer from the Giant Bonobo Diamond. It was the even brighter light that accompanied invisible zip-line travel. Blaine and Tracey found the SCIDAT data they needed for rain, monsoons, thunderstorms, and floods, which they had sent in with pictures to Uncle Cecil's basement.

After they exited the temple through a small passage located under the diamond's display case, Garfield T. Wellington the Forth had offered the gem to the indigenous pygmies. The content native Congo tribe declined the jewel so the diamond was now on its way to the British museum.

With their adventure in deepest darkest Africa completed, the Sassafrases had opened LINLOC and seen that their next

**LINLOC** SCIDAT
**LOCATION:** Patagonia
**CONTACT:** Hawk Talons
**LATITUDE:**    **LONGITUDE:**
-42° 00' 16"    -71° 27' 16"

**INFORMATION NEEDED ON:**
Snow, Ice storms,
Frost Quakes, Seasons

location was Patagonia, longitude -42° 00' 16" latitude -71° 27' 16". Their local expert's name was Hawk Talons. They would be studying snow, ice storms, frost quakes, and seasons.

Just reading the list of topics had left Blaine and Tracey shivering cold, especially after being wet and soaked virtually the entire time they had been in the Congo. As usual, though, they were looking forward to a new scientific adventure.

Jerk! The zip-lining motion stopped and the twins' bodies unclipped from the lines and slumped down, void of strength and sight. What usually happed next was that the blind white light would fade into color. Their strength would return as well at a similarly slow pace, but this landing was proving to be a little odd. Their sight should've returned by now, but everything remained completely white, and they could feel that their bodies hadn't completely stopped moving.

"Blaine, can you see anything?" Tracey shouted out. "Are we still moving?"

"No and yes," Blaine responded. "All I see is white, and we seem to be . . . dropping."

Both twins began reaching around with their hands and immediately felt snow.

"We must be in a snowstorm, Tracey," Blaine exclaimed. "That's probably why we can still only see white, but as to why we are dropping, I still haven't figured that out. At least we don't seem to be dropping that fast."

Tracey continued to feel around in the snow until her hands felt two things: splintery wood and an edge. "Blaine, I think we're on top of a big box; a big falling box."

A strong blustery wind whipped at the twins. Tracey continued to feel around the edge of the supposed box they were on, when she suddenly felt something else. It was some kind of strap or rope at the corner where two edges met. Immediately,

Tracey figured out where they had landed.

"Blaine! We are on top of a box that is connected to a parachute!"

Blaine felt around a little more and came to the conclusion that his sister was right. The invisible zip-lines had landed them on a wooden box, dropping through the sky by parachute in the middle of a snowstorm. They didn't have very long to wonder about their predicament because before the twins knew it, they were landing with a thud.

The collision of the box hitting the ground was enough to knock both Blaine and Tracey off their perch. The twins went sprawling. Luckily, they landed softly and harmlessly in a mound of fresh snow. Tracey immediately picked herself up, but Blaine took the time to make a snow angel.

"C'mon, Blaine, get up!" Tracey yelled, annoyed, even though she really wasn't. "We've got to figure out where we are and find our local expert."

"What's the rush, Trace? We're in a winter wonderland. Why don't you make a snow angel with me?"

Tracey shook her head no. "C'mon, Blaine, it's freezing. Let's get moving."

The Sassafras boy stood up to his feet and shook the snow off himself, almost like he was a dog.

"Hey, what do you think is inside the box?" Blaine asked his sister.

Tracey was about to offer a guess when suddenly an object landed in the snow off to their left. Whatever it was, it looked to be connected to a parachute just like their box. Then, to their right, something else landed. Then, right in front and behind them—objects seemed to be landing all over the place.

The snow was still coming down, but the twins noticed that

visibility was becoming steadily clearer. They could see that all the falling objects were connected to parachutes. They were smaller than the box was, and all the objects were . . . talking.

"It's people, Blaine!" Tracey tugged at her brother's arm.

"This is ridiculous!" One of the voices exclaimed in irritation. "I've had enough of this! Get me out of this thing!"

"Oh, come on, Ted. It's okay, man," another voice soothed. "This is the adventure of a lifetime! Haven't you always wanted to go pear-shooting out of a plane?"

"It's called parachuting, not pear-shooting. And no, I have never wanted to do this."

Now a woman's voice chimed in. "At least this is better than being in the office. And I've heard the survival expert is quite a hunk, right, Barbara?"

"Uh, yeah, I guess, Tammy," another woman's voice replied. "I jammed my finger just now when we landed. I'm in a little bit of pain here, but I will be all right."

The four silhouettes fumbled with their parachutes until they were unhooked. Then, they staggered through the snow toward Blaine and Tracey's spot.

"Oh, look, two kids have joined our adventure as well. Now, not only do we have to survive; we have to babysit," one of the men said sarcastically when he saw the twins.

The man brushed past the twins with a perturbed look on his face.

"Don't mind Ted." The next man smiled. "He's not really the adventurous type, and he's pretty much always grumpy."

"He sure is." A woman agreed as she joined the man.

Another woman joined them but she didn't say anything because she was too busy looking over her finger as if it was hurting.

"My name is Mitchell," the man offered cheerfully. "And this is Tammy and Barbara and of course you met Ted. We all work in the office together at Q. B. Cubicles. We sit in cubicles all day and sell cubicles—it can be somewhat monotonous. But just this morning, the four of us were kidnapped from our office, blindfolded, put in an airplane, and then dropped off here. We are excited because we think we are going to be on that TV show called 'Out of the Office.' Have the two of you heard of that show before?"

The twins shook their heads to say no.

"It's a show that teaches about survival and science," the woman named Tammy said. "And it's hosted by a complete studmuffin, but enough about us. Who are you two?"

Blaine began to respond, but he was interrupted by the approach of several more individuals. The man leading the pack was a big muscular man with a chiseled jaw outlined by a five-o-clock shadow. He was wearing the latest in outdoor apparel and looked completely in his element here in this cold weather. The group that flanked him was an entire camera crew, with all kinds of equipment and differing responsibilities.

With cameras rolling, the man walked right up to the twins and the group of office workers. "Welcome to 'Out of the Office,'" he said in a clear voice. "My name is Hawk Talons, and I will be your host, your scientist, and your survival guide."

"Hawk Talons?" Ted questioned, rolling his eyes. "Seriously? What kind of name is that?"

"It's the name they gave me when I was in Antarctica's Special Forces."

"I didn't know Antarctica had Special Forces," said Mitchell.

"They did," Talons confirmed. "It was me."

Ted gulped and rolled his eyes again at the same time.

"You have been kidnapped from your office and dropped here in the wilderness with just the clothes on your backs. Your goal is to find civilization, and in the process learn some useful survival skills. I will be with you every step of the way, pushing you, encouraging you, and teaching you the science of it all. You will face several official challenges and be given worthy rewards if those challenges are completed successfully. Does everyone understand?"

"I understand perfectly, Hawk," Tammy sang as she batted her eyes.

"I understand that this is ridiculous," Ted quipped. "Where in the world are we, anyway? I didn't ask to be kidnapped, and I don't want to be on this TV show. If you're such an expert, get us off this freezing mountaintop!"

"Very good observations, sir!" Hawk praised Ted. "It is freezing and we are on a mountaintop. To be even more precise, we are in Patagonia in southern Argentina. We are on top of one of the countless unnamed high altitude mountains. The northern portion of the Patagonia region is part of the pampas, aka grasslands. The southern part, however, transitions from the pampas to the taiga, so it is much cooler with the possibility of frost occurring throughout the year."

"Okay, never mind about our exact location," Ted interrupted. "Just get us out of this blizzard! My coworkers and I don't even have a coats on, for goodness sake!"

"Right again," Hawk responded. "We are experiencing blizzard-like conditions. Blizzards are long lasting snowstorms with strong winds. They can deposit a large amount of snow in a relatively short amount of time."

"Tell us more about snow," Tammy requested, evidently unaffected by the blustery conditions.

"Snow forms by a process of deposition," Hawk shared,

taking Tammy up on the request to share more. "This means that water vapor high in the atmosphere collects around a piece of dirt or dust and changes directly into ice without becoming a liquid first. The temperature must be quite a bit below freezing for this to occur.

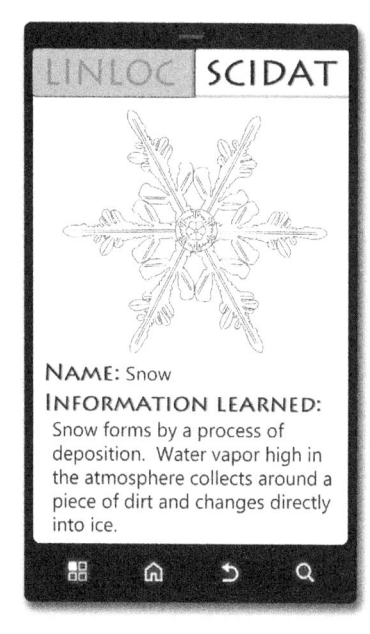

> **LINLOC SCIDAT**
>
> **NAME:** Snow
> **INFORMATION LEARNED:** Snow forms by a process of deposition. Water vapor high in the atmosphere collects around a piece of dirt and changes directly into ice.

"If snow meets any warm air as it falls to the ground, it can be turned into rain, sleet, or freezing rain. Also, snow is white because the crystalline structure reflects all light waves, making it appear white to our eyes. And, finally, snowflakes come in many shapes and sizes, but each one is six-sided. They form as many as two hundred individual ice crystals that come together in a lattice structure around a tiny piece of dust or dirt."

"Oh, oh! Can you tell us about the windshield factor?" Mitchell asked raising his hand.

Hawk smiled. "I believe you are referring to the wind chill factor. There is quite a bit of wind here in the Patagonia region, which can make it feet colder than it really is. This phenomenon is known as the wind chill factor. Wind chill is when the temperature your body feels is the air temperature plus the wind speed."

The twins took pictures of the falling snow with their phones as the survival expert paused briefly after giving his information.

Hawk took a deep breath, looked directly at the camera, and said, "Know the science of the earth. Know the science of survival."

"My, oh my, everything you just said was so amazing," Tammy swooned.

"It's not amazing, it's ridiculous," Ted complained. "Get us off this mountain, Mr. Talons, or whatever your name really is."

"It's too late for that," the survivalist said to Ted.

"What do you mean it's too late?"

"I mean that nightfall is quickly approaching, and if we try right now to get off the mountain, we will all freeze.

### *Icy Impositions*

"Our priority right now is to create shelter. Coincidentally, that need leads us to our first official 'Out of the Office' challenge." Hawk announced.

"Oh, man! This is so exciting," Mitchell gushed.

"Your first challenge is to build a snow cave," Hawk announced. "Each of you must dig out a hole big enough for at least two people to sleep in. The hole, or cave, that I deem the best will win the reward."

"Awesome! What's the reward?" Mitchell asked.

"The reward for this first round is a plastic water bottle."

Mitchell and Tammy clapped at this news, but Ted rolled his eyes.

"What kind of reward is a plastic water bottle? That is pathetic."

"In any survival situation, staying hydrated is of utmost importance," Hawk Talons answered. "So the water bottle is a worthy reward."

This made sense to everyone except Ted.

"One last thing before we start this challenge," Hawk continued. "As it is on every episode of Out of the Office, we have dropped in a mystery crate."

Mitchell and Tammy both clapped their hands again.

"Inside the crate are items that can help to aid you in survival. At first glance, the mystery items might not make any sense, but if you think outside of the box, they can prove to be invaluable. So go ahead, office staff from Q.B. Cubicles, go open your mystery crate!"

Blaine and Tracey joined the group of cubicle salesmen as they ran over and started opening the big wooden box that the twins had floated in on. It took the group quite a long time to pry the lid off and when they finally did, all their excitement ebbed away. None of them liked what they saw in the crate, and they had absolutely no idea how what they saw could aid them on the snowy mountaintop for survival purposes.

"You've got to be kidding me," Ted blurted. "Cubicles? You dropped in a big crate of cubicles for us?"

"We sure did," Talons answered. "And if you think creatively, these cubicles can be extremely useful every step of the way."

"That's impossible!" Ted said in disgust.

"Right now you must summon the strength and brains to change the impossible into the possible," the survival expert encouraged. "Because it is time to start the first challenge! When I say 'GO,' all six of you must attempt to dig the best snow cave. You have fifteen minutes. Are you ready?"

Everyone except Ted nodded their heads.

"GO!" Talons shouted.

Blaine, Tracey, and Tammy all immediately spread out, fell to their knees, and started digging furiously in the snow. Mitchell and Barbara started pulling cubicles out of the crate, trying to figure out how to use them. Ted just stood there, shaking his head in disgust. The camera crew spread out, attempting to get video footage of everyone and everything that was happening.

"Ouch! Oh, ouch!" Everyone heard Barbara cry out. "I just got a splinter, so I'm in a little bit of pain here, but I will be all right."

"How do you think we can use these cubicles, Barbara?" Mitchell asked his hurting office mate.

"Maybe take four or five of them and stack them like dominos?" he mused. "Maybe partition off a space to put a computer? Man, oh man, I don't have any idea at all how to use these cubicles out here."

Over where the three diggers were, there was a lot less talking and a lot more working. Blaine wasn't sure how Tracey or Tammy were doing, but he felt like he was making quick headway with his cave. He had already dug a hole the length of his body, and he was still going strong. He remembered from the earlier instructions that his cave needed to be able to sleep two people. As he dug, he was busy thinking about those cubicles. "How could a cubicle possibly help him out with his cave?" Blaine wondered as he quickened his pace.

Suddenly, it hit him. He knew what to do with the cubicle!

"Five minutes!" Blaine heard Hawk shout out. "Only five more minutes to finish your caves!"

The boy diligently worked on his cave until it was big enough for two adults to crawl into and lay down comfortably. Then, he ran over to the crate and pulled out a cubicle. Blaine grabbed the cloth material at the top corner and pulled down hard. The fabric came loose. Once he had pulled the entire piece off, Blaine raced back to his snow cave. He laid the large piece of cloth from the cubicle down, like it was a blanket. He then exited his cave and shouted, "Finished."

Hawk Talons came over to Blaine's cave with a cameraman following him. The survivalist peeked his head into the hole and smiled. He then looked at his watch and shouted out, "Thirty seconds, people! Only thirty more seconds to finish your caves!"

When the time allotted for the competition concluded,

Talons circled everyone up and addressed them. "Some of you took too much time messing around with the cubicles and barely got holes dug at all. Some of you got sizeable holes dug, but didn't even try to incorporate the cubicles. Only one of you dug a good cave and used material from the cubicle. So, for the first Out of the Office challenge we have a clear winner, and it is this young man right here."

Talons walked over to Blaine, patted him on the back, and handed him a cool 'Out of the Office' branded plastic water bottle. Everyone except Ted clapped for the Sassafras boy.

"Okay, nighttime is almost upon us. Everyone pair up, climb in a snow cave, wrap up in cubicle cloth, and try to get some sleep. We have a lot more in store for tomorrow, and if we want to find civilization, we need some rest." Hawk directed.

"This is ridiculous," Ted complained. "We can't sleep in snow caves! We will freeze to death."

"It is bound to be cold," Hawk confirmed. "But inside the caves, wrapped up in the material from the cubicles, sharing body heat, you will have enough warmth to survive. Know the science of the earth. Know the science of survival."

With that statement, the camera crew turned off the cameras and packed up the equipment. Then, they headed to some heated tents they had set up, effectively leaving the contestants out in the cold. The six reluctantly paired off and headed to their respective snow caves.

The Sassafras twins shared a cave, and surprisingly, they both slept great. The fact that they hadn't actually had a good night's sleep since they started studying earth science, coupled with the sonic lag, probably had something to do with that.

Before they knew it, it was morning. Blaine and Tracey now found themselves standing in a circle in the snow with Ted, Mitchell, Tammy, and Barbara being addressed again by the consummate Hawk Talons.

"Good morning, everyone! I see that you all survived the night in your snow caves. How is everyone feeling?"

"I think I may have a little frostbite on one of my toes," Barbara answered. "So I'm in a little bit of pain here, but I will be all right."

"I'm feeling absolutely wonderful," Tammy replied with a bright smile.

Ted just stood with his arms folded and a frown on his face.

Mitchell bent over and started scooping up some fresh snow and began putting it in his mouth.

"Hold on there, pal," Hawk said to the snow eating cubicle salesman. "Don't eat the snow."

"Why not? It wasn't yellow."

"No, that's not the reason at all. If you eat snow, it can lower your core body temperature and send you into hypothermia. If you ever get stuck in a place like this again, you should have a water bottle like the one Blaine won yesterday. The best thing to do is pack it with snow and then put inside your clothes close to your body and start moving around. Your body heat will melt the snow and turn it into suitable water for drinking. If you don't have a water bottle, you must find other ways to try and melt snow or get to lower altitudes and find other sources of water. And speaking of lower altitudes, that is our first goal today: to get down off this snowy mountaintop. Does anyone have any ideas on which way we should go?"

"This way!" Ted immediately started walking down a hill away from the group.

Hawk looked at those who were left. "Ideally, we want to head north toward the potential of warmer weather and possible vegetation. Ted must have luck on his side, because that is exactly the direction he is headed. Let's follow him."

The survivalist, the twins, the office workers, and the entire camera crew trudged through the snow behind Ted. There was still a bit of snow falling, but it was much less than the previous day, so visibility was much better. After a few minutes of walking, Ted suddenly stopped up in front of them.

He threw up his hands in the air and shouted, "This is ridiculous!"

He continued to repeat the phrase until the rest of the group got to where he was and they all saw the source of his frustration. Ted was standing at the edge of a sheer cliff, with no visible way down.

"There is no way off of this thing," Ted spewed out. "We are going to be stuck on top of this freezing mountain forever!"

"Or are we," Hank questioned.

The survivalist turned and faced the entire group. "This moment brings us to our second official 'Out of the Office' challenge!"

Excitement shot through the group.

"The challenge is to find a way to get down this cliff face. This time you will split up into groups of two and work together to find the best way down. Your team member will be the same person you shared a snow cave with last night, and remember, you can use the cubicles and anything else you have had in your possession since you were kidnapped from the office. Again, for this challenge, you have fifteen minutes. Is everyone ready?"

Everyone nodded, except Ted, who was shaking his head in disgust.

"Before you start, let me tell you that the reward for winning this challenge is a flint stick."

"Another lame reward," Ted said partially under his breath.

Talons ignored him. "A flint stick can provide you with fire if used correctly. Fire is another essential in most survival situations."

Hawk raised his hand over his head then shouted, "Go!"

Tammy and Barbara huddled together to start their brainstorming session. Ted and Mitchell got together, but Ted just stood with his arms folded as Mitchell thought out loud and acted like he was literally trying to pull ideas out of his brain. The Sassafras twins also got together, and immediately Tracey had a plan.

"We can't use our harnesses or our carabiners because that would lead to too many questions. However, I know what we can do," she shared with Blaine. "We can use the parachute cords as rope and we can rappel down the cliff!"

Blaine nodded as though he approved, but then he frowned. "But what are we going to attach the cords to? There's nothing but snow up here."

Tracey was frowning now, too. They needed to think this through a little more.

"We could use the cubicle to sled down, or, oh, oh, we could use them to make an airplane!" Mitchell was saying loudly.

"Mitchell! That is ridiculous! How on earth could we make an airplane out of a cubicle?"

Tammy and Barbara were being much quieter than Ted and Mitchell. By the looks on their faces, they were not coming up with any good ideas either. Tracey looked over at the stack of cubicles which had been dragged over by some of the crew members. Suddenly, the Sassafras girl was struck by another idea.

"Blaine, look!" she said, and then she led her brother over to where the cubicles were. "Look at the cubicle's feet!"

"Feet? What? Cubicles don't have feet."

"Sure they do. See those 'T' framed metal braces that stick

out of the bottom to help the cubicle stand upright."

"Oh, yeah, now I see what you're talking about. But how in the world are cubicle feet going to help get us down the cliff?"

"We can use them as an anchor," Tracey exclaimed.

Blaine gave his sister's new idea another nod of approval, this time with an added smile. The twins worked together to pull one of the cubicle feet off. Then they ran over and scavenged long lengths of cording from the parachutes, which were nearby as well.

"Five minutes," they heard Hawk Talons call out. "Only five more minutes remaining in this, the second competition!"

The Sassafrases hurried over to the cliff's edge. Tracey started tying together a make-shift harness with some of the parachute cord, while Blaine jammed the metal 'T' frame deep into the snow and made sure that it was secure. They got the hand-made harness tied to the cording, and the cording anchored to the 'T' frame just as a shout from Talons came.

"Thirty seconds! Only thirty seconds to finish!"

"Well, maybe we could make a slide or a trampoline?"

"No, no, no! That's ridiculous! Besides, how are either of those things going to help us?" Everyone heard Ted and Mitchell arguing.

"Time's up," Hawk shouted as he came over to look at what the Sassafrases had set up. The big man smiled.

"Again, I say, time's up!"

The two cubicle salesmen heard Hawk this time, so they came over with nothing at all to show for this competition. Tammy and Barbara joined the group, also void of any plan to get down the cliff face.

"I think it's pretty obvious who the winners of this competition are," Talons said, pulling out two 'Out of the Office' branded flint sticks. He tossed a stick to each of the twins.

THE SASSAFRAS SCIENCE ADVENTURES

"Great job, Blaine and Tracey!" he congratulated. "Not only are you the only team to come up with a plan to get down this cliff, you came up with a good plan. What you have set up here is the best thing you could have devised with the materials you were afforded. In some extreme rappelling situations, if the snow is the right consistency, it is possible to make an anchor just out of snow by digging a wide deep circular line in the snow and looping your rope or cording around that, but what you have done here with the 'T' frame has made this rappelling rope all the more secure. Well done, you two."

Everyone except Ted congratulated the twelve-year-olds. Then they all set up anchors and cords in the same fashion as the Sassafrases and began the rappel down the cliff. Ted complained the whole way down. Tammy laughed in delight. Barbara smacked her knee against rock during her descent and shouted, "I smacked my knee so I'm in a little bit of pain here, but I will be all right." Mitchell questioned why this was called 'red-pailing' if there was no red pail incorporated in the activity at all.

Blaine and Tracey, however, just slid down in silence, jumping out from the rock face and then swinging in, landing on and pushing out with their feet. They were enjoying the entire experience.

The 'Out of the Office' cameramen caught everyone's descents on film, rappelling down next to each individual using real ropes and harnesses. Some of the crew lowered down what was left of the parachutes and cubicles as well.

When everyone and everything was down safely, Hawk Talons gave the directive to head toward a distant tree line to the north. As the group headed that way, the conditions worsened. Now, when the snow hit them in the face it hurt. The snow seemed to be harder and more solid down here off the mountaintop.

"Oh! Ouch!" Barbara cried out. "The end of my nose just got pinged really hard with a chunk of snow, so I'm in a little bit of pain here, but I will be all right."

"It's not snow. We're in an ice storm," Hawk Talons hollered. "We need to get to those trees as quickly as we can."

As the group quickened its pace, Talons continued to share, "An ice storm is a winter storm with sleet, or ice pellets, instead of snow. As we descended down that cliff face, did you feel that pocket of warm air? It was the cause of the change in the conditions. You see, ice storms occur when there is a layer of warm air sandwiched by two layers of cold air. The snow falls, melts, and the refreezes as ice as it reaches the surface. Not only did we need the layer of warm air for this ice storm to occur, we also needed the air close to the surface of the earth to still be below freezing, which is what the snow-covered ground has provided.

> **NAME:** Ice Storm
> **INFORMATION LEARNED:**
> Ice storms occur when there is a layer of warm air sandwiched by two layers of cold air.

"The ice pellets don't typically stick to surfaces, but sleet can accumulate just like snow. Know the science of the earth. Know the science of survival."

With that, the survivalist started for the trees. Everyone in the group, including the camera crew, followed suit. They seemed to be running directly into the storm. Freezing rain lashed and whipped at them, making any piece of bare skin pay the price. Blaine and Tracey both somehow managed to snap pictures of the weather as they battled forward.

The wind was blowing so hard that hats and hoods were flying off of people. Blaine looked to his left and watched as one of the cameramen lost the big fluffy stocking cap that he was wearing.

The man then looked in Blaine's direction. The wind was strong and the weather was thick, but it looked to Blaine like the man didn't have any eyebrows.

## Chapter 7: Out of the Office

### *Frost Quake!*

The trees definitely offered some needed shelter, and the storm seemed to have slowed down quite a bit, but the ground was still white with snow and freezing rain. Everyone was cold and wet.

"How is everyone holding up?" Hawk Talons asked the group, as everybody tried to find the driest places possible to sit down.

"We are miserable!" Ted answered on behalf of the group. "This is ridiculous! You have us trouncing through sleet and snow! You have us sleeping in caves and hanging off of cliffs! I don't want to be a part of this anymore! I don't want to be on this TV show! For the love of Pedro, I am still in my office clothes!"

"You're right," Talons agreed. "You are still in your office clothes, but do you have the will to survive?"

"What?" Ted asked with exclamation.

"Do you have the will to survive? Do you have in your heart and soul what it takes to get out of your current predicament alive? Can you find the strength to press on, survive, and eventually find civilization?"

Ted didn't answer, but for the first time, he had a look on his face that didn't include a frown.

"I know you're still in your office clothes," Talons repeated. "Ideally, in a cold and wet setting like you are in now, you would be wearing several layers of wicking or quick drying clothes, not made of cotton. Cotton absorbs water and stays wet longer, which can usher in hypothermia. Other materials such as wool or polyester will keep a person drier and therefore warmer. But that is not how any of you are dressed right now, and there are no retailers in sight. So, what are your options? How will you survive?"

"Make a fire," Tammy shouted out, still obtaining most of her optimism and all of her crush.

"That's absolutely right!" Hawk smiled. "And that leads us to our third official 'Out of the Office' challenge!"

The group would have clapped again, but their hands were all feeling too frozen.

"This will be an individual challenge with simple parameters. The first person to start a fire with a flame that is sustainable, wins. The reward for this challenge is a small mystery container," Talons said, holding up a small circular tin.

"Obviously, you can't see what is inside, but trust me—it is a worthy prize. Hopefully, that piques your interest enough to fight to win," Hawk added, with a knowing smile.

"And, might I just say, Barbara, Tammy, Mitchell, and Ted, you need to get on the ball in these challenges. So far, only the twelve-year-olds have won."

"And, one more surprise before you start..." The survivalist reached into a bag at his side. "Each of you will have the use of a flint stick for this challenge."

He tossed a stick to the four QB Cubicles employees.

"These are smaller than the ones Blaine and Tracey got, but they should do the trick. Is everyone ready to start?"

Everyone nodded, mostly because they were ready to be warm, not necessarily because they wanted to compete in another challenge.

"You will have fifteen minutes for this challenge. Remember to keep the cubicles in mind. Ready . . . Set . . . Go!"

Tammy and Barbara both hopped up and ran out to forage for any sticks and branches they could find. Mitchell and Ted both ran over to where the cubicles and been stacked to start the challenge there.

Blaine and Tracey, however, remained seated where they were. They weren't really paying attention to the current competition. Instead, they were just looking at different members of the crew. Blaine had told his sister that he thought he had seen the Man with No Eyebrows posing as a member of the 'Out of the Office' crew. Since that point, all the Sassafras twins seemed to be able to do was stare at the TV show-making squad.

"Blaine and Tracey, aren't you two going to tackle the fire challenge?" Hawk Talons's voice said, breaking off the twelve-year-olds' stares.

"Oh, uh, yes," Blaine blurted as he and Tracey got up and went in different directions, looking for something to burn.

"Ouch! A dead branch just fell and hit me in the head, so I'm in a little bit of pain here, but I will be all right," Barbara shouted out as she proceeded to drag the branch away.

At the same time, Mitchell could be heard shouting out as

well. "Hey, the inside part of the cubicles are wooden!" He was looking at a cubicle that had its cloth covering ripped off.

"And wood is flammable, right? So maybe I can catch a cubicle on fire!" The salesman then proceeded to strike his flint stick against the broad side of the bare, upright cubicle.

Tammy had now returned with a sizeable armful of twigs and branches. She dumped her pile of wood on the ground and started trying to get it lit.

Ted was near Mitchell, looking at a bare cubicle as well. He was kneeling down and picking at the wood at one of its corners.

Blaine and Tracey were in different spots, but they were both just kind of walking aimlessly through the trees, half looking for firewood and half looking for the Man with No Eyebrows.

Blaine began wondering if maybe he had just imagined seeing the eyebrowless villain. He was pretty sure he had spied out each crew member, and all of them had eyebrows. They all looked to be legitimately working. He knew that the Man with No Eyebrows had access to the invisible zip-lines, so he could appear and disappear just like they could. Blaine also knew the Man with No Eyebrows was relentlessly trying to thwart his and Tracey's science learning. He was always popping up and wreaking havoc. The twins just didn't know what this mysterious man's motivation was.

On their botany leg, Blaine had actually caught the Man with No Eyebrows and they had wrestled around. Blaine had hoped to stop him or at least get some answers. But the man had used the invisible zip-lines and the Dark Cape, which was a magical disappearing suit he had stolen during their anatomy leg, to escape yet again. The twins still really had no idea who this annoying villain was.

"Five minutes! Five minutes left to finish your fires!" Hawk Talons's voice sounded out, causing Blaine's mind to switch gears.

Blaine looked at all of the branches in his arms, which he had absent-mindedly picked up. "This is plenty. Now I just need to get it lit," the boy thought.

Tracey was already down on her knees next to her pile of wood. She knew you had to get the small pieces of wood lit first and then use slightly bigger pieces to build a kind of teepee around the smaller pieces. But every time she had seen her dad build a fire, they had used matches or a lighter, not flint. She began striking her flint stick and immediately saw sparks. Tracey smiled—she was optimistic that she could figure this thing out.

Mitchell was still standing over by his upright cubicle striking his flint stick feverishly against the big bare piece of wood, but to no avail.

Tammy had not gotten her pile of wood lit. Barbara had not gotten her branch lit, but surprisingly, Ted had gotten a small fire started. Instead of complaining, he was gently blowing underneath the small flickering flame to grow it and carefully adding tiny branches to fuel it. Mitchell, Tammy, Barbara, and the twins continued to work at getting their fires started, but it was pretty clear who was going to be the winner.

"Thirty seconds! Only thirty seconds to finish!" Talons announced loudly.

Sparks flew. People blew. But only one sustainable flame appeared.

"And . . . stop! Everyone step back," Hawk directed.

With that, everyone threw down their hands, pocketed their flint sticks, and stepped back.

"We have a new winner!" Hawk gestured towards Ted and his crackling fire.

"Well done, Ted! And let me share how Ted was able to start a fire when none of the rest of you did. In a setting like we are in now, one that is snowy and wet, it is very difficult to find anything dry enough to burn.  Blaine, Tracey, and Tammy, that was the problem the three of you ran into.  You each had fairly small pieces of wood with which to start your fires, but it was all just too damp.  You need both small and dry kindling when starting a fire, especially when you are using flint.

"Mitchell and Barbara, the pieces of wood that each of you were trying to light were just too big.  Ted, however, saw that the cubicle interiors were made of particleboard, which contains small pieces of wood bonded together.  So he pulled and scratched at a corner of the particleboard until he was able to get a whole handful of small, dry, almost dust-like pieces of wood.  Then, he dug a small hole and put a rock at the bottom of it, again trying to create the driest environment possible.  Finally, Ted made a pile on the rock with the particleboard pieces and after a dozen or so strikes, a spark from his flint stick caught the pile of kindling on fire.  He then added the smallest, driest twigs he could find.  And as his flame grew and got healthier, he added progressively bigger pieces of wood."

Hawk paused before looking into the camera to say, "Know the science of the earth.  Know the science of survival."

He then turned toward Ted and patted him on the back. "Well done, Ted, and here is your reward for winning the fire-making challenge—the mystery container."

The survivalist handed the victorious cubicle salesman the small tin can.  Ted immediately opened up the can with a tinge of excitement, but his face was soon overtaken by a scowl.

"What kind of ridiculous reward is this?  It's a container full of dust!"

"That's not dust, my friend," Hawk retorted.  "That is seasoning!"

"Seasoning?" Ted questioned.  "How in the world am I going to use seasoning out here?  We don't have any meat to add seasoning to! As a matter of fact, we don't have any food at all! This mystery container stinks!"

Not paying any attention to Ted's complaints, Hawk Talons changed the subject. "Everyone spend some time getting dry and warmed up around the fire, and then we will press on in our quest to survive and find civilization."

A good while later, the entire group was on the move again. As they proceeded, it was apparent they were getting lower in elevation. They spent hour after hour weaving around trees and other growth until they eventually got to a large rock-surfaced field.

Blaine and Tracey had both diligently studied every single member of the crew as they hiked, but they didn't see one person without eyebrows. The twins were beginning to be convinced that Blaine had just imagined seeing their mysterious antagonist.

Another tree-line appeared on the other side of the field in which they were walking. According to Hawk, that is where they were headed. As they walked towards the trees, the Sassafrases noticed that the snow and sleet seemed to be melting.

The field ended up being much bigger than it had originally looked, and by the time the group reached the trees, everyone was feeling spent and depleted. It had been a long day, and they had exerted a lot of physical energy, and no one had eaten a thing.

"There will be no 'Out of the Office' challenge this evening," Hawk said, sliding his bag off and dropping it to the ground. "Instead, I will show you how to make simple A-frame tents using the parachute cording and the material from the cubicles."

This seemed to come as good news to everyone, as they fell, exhausted, onto the ground to watch the survivalist ply his trade. He took a length of cord and tied each end to two trees that were about ten feet apart. He then draped a large piece of the cubicle material over the line of cording, which was about three feet off the ground. Next, he pulled both sides of the material that were hanging down equally on either side of the cording, down to the ground and secured the ends to the ground by piling rocks on the material's edges. This created a sort of 'A' shape with the cording being the high point and the rock covered edges of the material being the two low points. Hawk then pulled off some small evergreen branches and made a somewhat soft and dry place to lie down on underneath the A-frame tent he had just created.

It wasn't long before night had fallen and the twins and cubicle salesmen all fell asleep and were snoring in their designated pairs in A-frame tents. The Out of the Office crew all slept soundly in their heated tents and comfortable sleeping bags.

Boom! Rip! Crack! Shake!

Shattering sounds rang out with fury, abruptly destroying the silence of the night. The ground gently rumbled and shook. The whole group awoke and shot up out of their tents.

"Ridiculous! Just ridiculous!" Ted screamed out.

The twins' first thought was a collective, "The Man with No Eyebrows!"

Tammy ran over and sidled up next to Hawk. Barbara attempted to do the same, but instead she tripped over a rock.

"Ouch! Oh, ouch! I've bumped my elbow, so I'm in a little bit of pain here, but I will be all right."

The shaking and piercing sounds didn't last long at all, only seconds, but it was plenty to capture everyone's heartbeats. Everybody was rubbing their eyes and looking around, trying to figure out what in the world had just happened. Strange lights could be seen rippling in the night sky, adding to the eeriness and mystery of their current setting.

"Frost quake," Talons finally said.

"Frost cake!" Mitchell blurted out only half awake. "Oh,

man, that sounds delicious, doesn't it, Mr. Rubber Ducky. Please don't hit me over the head with your spaghetti fan."

The group looked at each other and chuckled. It was obvious to everybody except Mitchell himself still partially immersed in quite the crazy dream. "Not frost cake—frost quake," the survivalist corrected, with a wide grin. "Also known as cryoseisms, frost quakes happen in the middle of the night when the temperatures are at their coldest. They are not caused by a shift of the earth's plates. Rather, they are caused by a sudden rapid freezing of the ground or by glacial movement.

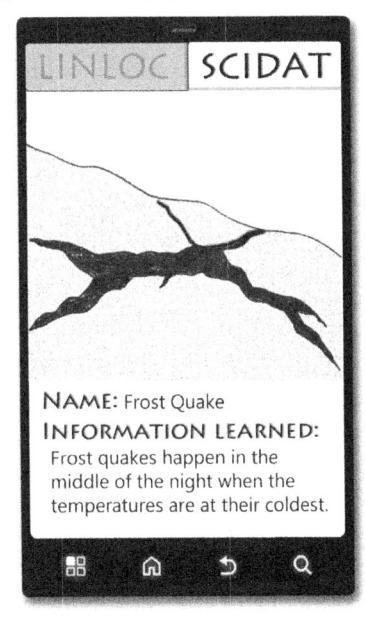

"For a frost quake to occur, water needs to saturate the soil and fill the gaps in the rocks of the glacier. Usually this happens when the temperature goes above freezing and some of the surface snow or ice melts. Then, the temperature needs to plummet quickly below freezing again. This causes the water to rapidly freeze and expand. Pressure is exerted on the surrounding soil and rock, forcing it to crack in an explosive manner, which creates a loud sound and can even shake the ground."

"So, that's what we just experienced?" Tammy asked. "That's what woke us up?"

Talons nodded his head. "Frost also happens at night. As the temperature cools, the water vapor in the air close to the ground condenses, freezes, and then forms ice crystals.

Hawk shone his flashlight out into the nearby field, where the twins quickly noticed a sparkling appearance caused by the frost. "Frost appears on the ground and other surfaces when the temperature falls below freezing. If the temperature does not go below freezing, dew forms instead of frost."

"What about those weird lights floating around in the sky? What are they?" Blaine asked.

"Those are the Southern Lights," Hawk said, gazing up into the sky almost dreamily. "Their official name is the Aurora Australias. These lights are an astronomical phenomena that look like shafts or curtains of light that dance in the sky during the winter months. They can also be found in the northern hemisphere, although there they are called the Aurora Borealis or the 'Northern Lights.' Of course, right now it is summer in the northern hemisphere, while it is winter in the southern hemisphere. So we get to see these lights right now, while they cannot see them up north."

Everyone nodded their heads as they just stared at the mysterious moving lights in the sky. They were like nothing the twins had ever seen, and they were quite hypnotizing.

Everyone forgot for a second how cold it was there on that Patagonian night, and they all just gazed at the winter wonderland that was before them. Frost sparkled over the expanse of the ground and the Southern Lights continued their rhythmic dance in the sky.

Blaine and Tracey snapped pictures of the frost with their smartphones before they returned to their A-frame. They had both all but forgotten about the possible presence of the Man with No Eyebrows.

Tammy and Barbara retired to their A-frame, as Ted led the half-awake Mitchell back to their tent. Hawk made sure all the contestants were safe in their tents before he joined the TV show crew as they all snuggled back into their cozy mobile abodes.

Thankfully, the rest of the night was silent and free of any frost quakes.

## *Seasonal Shifts*

In the morning, everyone was awakened by Hawk Talons's strong voice.

"Good morning! Good morning! Good morning! Who is ready for the fourth official 'Out of the Office' challenge?"

"Blah, blah, blah," mouthed Ted as he sat up in his tent. "Just the thought of another challenge at this point is ridiculous."

"Okay, then let me ask it this way," Hawk responded. "Who is hungry?"

Everyone shot up out of their tents at that question. The cubicle salesmen and the twins were all famished, and they were very ready to eat something . . . anything."

"It's been nearly a day since any of you have eaten," Talons informed them. "Our goal is to find civilization, but we won't have enough energy to accomplish that goal if we don't get some fuel in our bodies. You burned numerous calories getting from the snowy mountaintop to where we are now, and you need to replace those calories. It is vital that we find something to eat this morning."

"Spaghetti, sounds good," blurted Mitchell. "Do you have any spaghetti lying around . . . why does spaghetti sound so good?"

"We don't have any spaghetti," the survivalist responded, smiling. "But these woods are teeming with small animals that we can hunt or trap and then eat."

"Is that our next challenge?" Blaine asked excitedly.

"Yes it is, Blaine," Hawk answered. "Your fourth official 'Out of the Office' challenge is to come up with the best way to use the cubicles and/or the parachutes to capture a small animal we can eat. You will have an entire hour for this challenge. The person that comes up with the best idea will win a worthy reward."

"What's the reward this time, Hawkie?" Tammy asked like she hoped the reward was a date with Talons himself.

Hawk raised a skeptical eyebrow at being called 'Hawkie.' Then he answered. "The reward for winning this challenge is an 'Out of the Office' survival grade pocket knife."

Everyone in the group, including Ted, responded favorably to this news.

"Is everyone ready?" the big survivalist asked.

Everyone nodded.

"GO!" Hawk shouted.

The six 'Out of the Office' contestants scattered off in different directions, fueled in this challenge by their desire to eat. Mitchell ran off into the woods muttering something about making a fan to capture a ducky. Evidently, he still wasn't coherent enough to discern dream from reality. Ted immediately began climbing a tree.

Barbara began stacking up a pile of rocks which she promptly toppled over onto her left foot. "Ouch. Oh, ouch! I smashed my foot on a rock, so I'm in a little bit of pain here, but I will be all right."

Tammy stayed within an arm's length of Hawk Talons. She must have been hoping that he would feel sorry for her and divulge some hints on how to capture a meal.

The twins were really the only two to run over to the material piles. Blaine focused his attention on the parachutes, while Tracey studied the cubicles. Blaine made up his mind a little quicker than his sister as to what his plan of action was going to be. The boy grabbed some of the parachute cording and scampered off into the trees.

Tracey looked at the office partitions a little while longer and then reached over and pulled a long piece of the cubicle's aluminum frame off. The girl now basically had a six foot metal stick in her hands. Tracey ran over to the campfire, which the crew had started earlier in the morning, and plunged one end of the aluminum frame deep into the burning coals. She waited until she thought the lightweight metal would be red hot and bendable, before she pulled it out of the fire. Sure enough, the end of her aluminum stick was now glowing red. Tracey set that end down on a rock, grabbed another rock, and started carefully beating the end into a sharp point.

The Sassafras girl managed to bang out her desired shape. With a smile on her face, she held up her handmade spear. Tracey then ran into the forest in a different direction than anyone else had gone. She wanted to get away from the sounds of human chatter, so she could lie silently in wait and use her spear on an unsuspecting animal. An "Out of the Office' cameraman followed her, but other than that, Tracey was alone. The girl could tell that the lone hooded cameraman was attempting to be quiet as he followed her through the forest, and for that she was thankful.

After a five to ten minute hike, the Sassafras girl found a spot near a small stream that she thought was suitable, so she stopped and crouched there, with the cameraman off to her right just a little. Out in the distance, the girl heard a faint call coming from Hawk Talons.

"Thirty minutes! Thirty minutes left in the challenge!"

Tracey had gone deer hunting with her dad one time, and they waited in the deer stand three hours before they saw a deer. The Sassafras girl was hoping that she would see an animal much quicker than that this time. She figured animals probably frequented this little stream. She was sure this was a good spot to spy and wait, giving her a chance to use her makeshift spear.

Tracey did her best to stay quiet and still. Time began to slowly tick by with absolutely no sign of animal activity at all. Tracey was in a sort of crouched position. She hoped that she could hold this stance without any of her limbs falling asleep. The silent cameraman kept his camera trained directly on her. Tracey felt bad for him, that he had pulled the duty to be the one to follow her out here. This was probably pretty boring footage he was getting right now.

She looked at him and gave him a kind of apologetic smile as if to say: sorry this is turning out to be so lame. He didn't respond in any way at all. He just kept aiming his camera at her, with one eye looking into the lens and one eye staring, completely unblinking, her way. She couldn't see much of his face because his stocking cap was pulled down low, and his coat was pulled up high. She could basically only see that one eye.

A sudden thought hit Tracey. What if this cameraman was actually the Man with No—wait! What was that she saw out of the corner of her eye? Tracey saw something moving next to the stream. She turned her attention away from the cameraman toward the movement. Right there, about four feet down the stream to her left a fish splashed.

Tracey slowly crept to that part of the stream. She held her spear up, just like Arrio had shown her on the botany leg about a week ago. Tracey tightened her grip on the spear.

She knew she was only going to have one chance at this. In the course of about one second, it all happened. Tracey aimed, threw the spear, and got the fish directly in its center. She had done it!

She had successfully executed her plan. Suddenly the cameraman burst up from his quiet spot and ran toward the Sassafras girl. A chill of fear filled Tracey at the sudden movement, but the chill was short lived.

"Great job, Tracey," the cameraman said as he ripped his hood off in elation, revealing a pair of perfectly normal brown eyebrows. "You did it! You caught a huge fish! And I bet you won the challenge! Let's hurry back to camp and show Hawk!"

Half an hour later, after the challenge was officially over and everyone had been rounded up, Tracey found herself standing in front of everybody, with Hawk Talons raising up her arm up like a victor.

"Tracey was not only the first contestant to catch breakfast, she was the only contestant to catch a meal," the survival expert was saying.

"Blaine, good idea making the tree spring noose trap using parachute cording and a sapling. I bet you would have caught something if the challenge had been longer. Ted, your idea was similar to Blaine's, but it wasn't executed as well. You were running with a decent idea, but snares work far better on the ground than they do in the tree-tops. And the other three of you, what can I say other than you would be starving if it weren't for Tracey."

Talons proceeded to give Tracey her reward for winning the challenge. Then, he showed everyone the proper way to gut, clean, and prepare a fish to eat as a meal.

Even though Tracey was hungry, she was more than happy to share her catch with the other five contestants. She wasn't real keen on eating plain fish, so she was thankful that Ted had agreed to share some of the seasoning he won. When Ted finally took his first bite, he wasn't complaining anymore about his 'container full of dust' either.

The 'Out of the Office' crew sat on the outskirts of the six contestants' fireside circle and had their meal of delicious looking peanut butter and jelly sandwiches, chips, and candy bars. It was a little hard for the twins and the cubicle salesmen not to be jealous, but at least they were replenishing their bodies with calories.

"Besides providing a meal today, Tracey also got us one step closer to civilization," Hawk Talons announced when everyone was done eating.

"I did? How?" the Sassafras girl asked.

"You sure did," Hawk confirmed. "When you found the stream, you not only found a place where animals drink and live, you also basically found a route for us to use to get out of here. In most survival situations, if you can find running water, you can follow that water downstream where it will almost always eventually lead to civilization."

"Let's get going, then!" Ted said, standing up. "Let's follow that stream and let's get out of this ridiculous place!"

Hawk Talons looked at Ted and nodded. Then, he looked directly into a camera. "Know the science of the earth. Know the science of survival. Let's go!"

The whole group went together to find Tracey's stream. Then, led by the confident survivalist, they began hiking down, down, down, following the running water. As they went down in elevation, the stream continued to get bigger. The sun actually came out as they made their trek down, and when it did, it seemed to put cheer in everyone's spirits—even Ted's.

"Ahhh, nice," the cubicle salesman said, closing his eyes and turning his face toward the sun. "It's like we've had a changing of seasons!"

"It is nice, isn't it," Talons responded. "Winter is the longest season in Patagonia. Spring, summer, and fall are relatively short here, so it is always nice to get some sunshine. I know in the last couple of days we have faced a lot of snow and freezing rain, but Patagonia usually remains fairly dry, even though its weather is mainly influenced by the humid winds that blow off the Pacific. As these winds blow over the Andes Mountains, they lose most of their humidity through cooling and condensation. By the time they reach Patagonia, most of the moisture is gone."

"And what was it you said last night about the different hemispheres?" Ted inquired pleasantly.

Blaine and Tracey looked at each other, shocked. They couldn't believe the usually grumpy salesman was asking questions and was actually capable of such civility.

**NAME:** Seasons
**INFORMATION LEARNED:**
There are four seasons: winter, spring, summer, and fall, otherwise known as autumn.

"When it is winter here in the southern Hemisphere, it is summer in the Northern Hemisphere," Hawk responded to Ted's question. "Since the earth is slightly tilted on its axis, it tilts one hemisphere toward the sun, making the weather warmer, while the other hemisphere is tilted away from the sun, making the weather colder. That is why when it is winter here, it is summer up in the Northern Hemisphere.

"All in all, we have four seasons: winter, spring, summer, and fall, otherwise known as autumn. A season is a collection of days with a typical weather pattern. All these seasons are a result of the changing position of the earth as it rotates around the sun and the tilt of the earth on its axis. The earth's orbit around the sun is slightly elliptical, so during certain times of the year, the earth is either closer to the sun or farther from the sun than others, affecting warmth and frigidness."

Ted smiled as the survivalist finished giving information. "Know the science of the earth. Know the science of survival," the cubicle salesman said.

Though everything had been rather severely charred during the explosion, he had managed to successfully clean up and reuse almost every component. He smiled a wicked smile. This machine of his would be up and running in no time. The Forget-O-Nator would live to see another day!

He allowed himself to sit down and rest for a moment. He would get back to work soon enough, putting the last pieces in place and completing the finishing touches.

He smiled even bigger now, causing his forehead to wrinkle up where his eyebrows would have been, if he had any. When his work here was done, he could get back to one of his more enjoyable tasks: attempting to terrify and kidnap those Sassafras twins.

## Chapter 8: The Gobi Desert
### *Is it Day or Night?*

They had found civilization.

Then, they had zipped. And now all they found was . . . nothing.

The Sassafras twins just stood staring at the nothingness all around them. They could barely tell the difference between sky and land. There were no plants, no mountains, no clouds—just dust, flat ground, big sky, and nothing else.

Patagonia had been pretty barren, but it was like an arboretum compared to this place. The Sassafras twins, the Q.B. cubicle salesmen, Hawk Talons, and the rest of the 'Out of the Office' crew had followed the stream straight down to a little village. Finding the village had victoriously concluded their challenge to survive the wilds of Patagonia.

Blaine and Tracey had taken the time to send in all their SCIDAT data and pictures to Uncle Cecil's basement. Then, they had opened up the LINLOC application on their phones to see where they'd be going next. Mongolia had been the location that came up on their screens, longitude 104° 53' 00", latitude 42° 19' 57". They would be studying the topics of day and night, sandstorms, drought, and oases. Their local expert's name was Ganzorig Buri.

The twelve-year-olds had excitedly put on their helmets and harnesses. With great anticipation, they had calibrated their three-ringed carabiners to the correct coordinates. Within seconds, they had zipped off at top speeds.

They had landed safely and had quickly regained their strength and sight, but now they were confused. Where had they landed? This place looked like another planet.

"It was longitude 104° 53' 00", latitude 42° 19' 57", wasn't it?" Blaine asked. "And that's what we both turned the rings to, right?"

"I think so," Tracey answered blankly.

Blaine continued, "LINLOC is designed to land us as close to our local expert as possible without being detected, but there is no local expert here. There is nothing here."

The Sassafrases both turned in a slow circle and took a long stare at their new location. There was so much nothingness, it almost made the twins feel like they were going to float off into the void expanse, never to be heard from or seen again. They knew this couldn't really happen, but it was a scary feeling, nonetheless.

"Maybe this is like Alaska," Tracey offered after a few moments of silent staring.

"What?" Blaine asked. "What do you mean?"

"I mean maybe it's like Summer Beach's lab. Remember how we landed in that big field with nothing around, and then suddenly a hole in the earth opened up and we slid down the slide into the lab?"

Blaine looked around at the ground. Maybe this location was like Alaska. Maybe they were going to get to slide down into some kind of underground science lab. The twins stood waiting for something to happen. They waited and waited, but nothing happened.

"Maybe the old glitch in the zip-lining system that was fixed has been replaced by some kind of new glitch," Tracey said. "Maybe we should call Uncle Cecil to see what is going on."

Blaine pulled out his smartphone and began to do just that. But before he could push a button, Tracey put a hand on his arm and then pointed out into the blue-ish nothingness with her other hand.

"Look!" She said. "I see something! I see a dot on the horizon!"

Blaine put his phone down and looked. At first he didn't see anything because he couldn't even hardly tell where the horizon was. But as he continued to strain his eyes, he saw it. Tracey was right. There was a dot out on the horizon, and the dot was moving. It looked to be bobbing up and down and as the twins stared at it, and it seemed to be steadily getting bigger.

The image would bob, move, and then occasionally almost disappear in all the blue. Then it would reappear quickly, almost like it had been hidden briefly by the wind. It continued to get bigger and bigger, and the image of the dot soon began morphing into a clearer form.

"I think it's a person," Tracey exclaimed. "And it looks like he is coming our way!"

Blaine nodded, agreeing with his sister's summation, but he was pretty sure it wasn't just a person. "I think it's a person riding a horse," he added.

Whoever it was, he was coming straight toward the twins, and he was coming fast. After a few more moments, both children could see very clearly that it was indeed a person on a horse. A tinge of fear popped up in the Sassafrases hearts. What if the rider coming their way meant harm? There was nowhere to run and hide, and at this point, they didn't think they had enough time to calibrate their 3-ringed carabiners and zip away.

So the twelve-year-olds just stood in their places and anxiously waited to see what would happen. The horse and rider sped right up to the twins' spot and then stopped in a swirl of blue dust. The horse snorted, and the rider, who was a little taller than the twins but much stronger looking than they were, jumped down to the ground. The moment his boots hit the earth, a wide smile formed on his face, and he held out his hand for a shake.

"My name is Ganzorig Buri," he greeted heartily. "Welcome to the Mongolian desert."

Blaine reached out to shake the man's hand, but when he did, Ganzorig reached past Blaine's hand and grasped the boy's entire forearm for a much more manly greeting. As Ganzorig shook his arm, Blaine was sure he had never felt a tighter or stronger grip in all his life.

"We are the Sassafrases," Blaine said. "Blaine and Tracey Sassafras."

"Very nice to meet you, Blaine and Traccy Sassafras," the Mongolian man said, as he let go of the Sassafras boy's forearm.

Ganzorig looked ahead in the direction he had been riding, and then turned and looked back to where he had come from. "I am returning home from college," he stated.

That was the last thing the Sassafrases expected their new local expert to say. First off, he looked a little too old to be a college student. He was rugged and earthy and was wearing what the Sassafrases assumed to be traditional Mongolian clothing. Plus, the most obvious issue to the twins was the fact that they were in the middle of a desert with no visible landmarks in any direction. There seemed to be no possibility of a college or a home to return to; just miles and miles of nothing. But the Sassafrases believed their new local expert. They knew by now that appearances of both people and places could be deceiving.

"The Mongolian Desert is also known as the Gobi Desert," Ganzorig informed. "It is known as the land of the blue sky because clouds are rarely seen here."

The twins took another look around at their new location. "Land of the Blue Sky" was definitely a good name for this place.

"The climate in the Gobi Desert is very extreme," Buri continued. "It can reach over one hundred degrees Fahrenheit in the summer and negative one hundred in the winter. It is known as a continental climate, because it is not influenced by a neighboring sea. It is too far inland."

"It doesn't feel too hot or too cold right now," Tracey commented.

"You are right, Tracey," Ganzorig agreed. "It's still fairly early in the morning so the weather is pleasant. Rest assured though, it will get pretty hot as the day goes on. Here in the Gobi, every day ends up being pretty much the same."

Blaine looked around at the morning sky, trying to figure out exactly how many days he and Tracey had been zipping through their earth science studies. All the days pretty much seemed to run together.

They had started in Oklahoma City and had spent a whole day there without sleeping. Then they had zipped to the next day in the Congo and also spent a day there without sleep. They then zipped back to the morning of that same day in Patagonia. There at the frozen tip of South America, they had finally gotten a couple nights of sleep, but both of those nights had been cold! He and Tracey had zipped ahead again another day and had landed in the morning here in the Mongolian Desert. At least that's how Blaine thought it all had happened . . .

"As the earth turns or rotates, day changes to night on one side of the globe and on the other side, night changes to day," Buri said, almost like he was addressing the unspoken questions in Blaine's mind.

"This is because, as the earth turns, different parts of the surface face the sun. In the morning, our side of the earth is turning to face the sun. This is when we see the sunrise, which ushers in a new day. In the evening, our side of the earth is rotating away from the sun. This is when we see the sunset, which marks the beginning of a new night. The day/night cycle takes a full twenty-four hours to complete."

**SCIDAT**

**NAME:** Day and Night
**INFORMATION LEARNED:** The day/night cycle takes a full twenty-four hours to complete.

Blaine was still confused. This was all good SCIDAT info he and Tracey could send to Uncle Cecil, but it didn't really help at all in figuring out exactly what day it was. Or how long he and his sister had been zipping through earth science.

Tracey, however, was fully engaged in the science of what

Ganzorig Buri was saying. "So, would this desert be considered a hot or cold desert?" the Sassafras girl asked.

"The Gobi is a cold desert," Buri answered. "Cold deserts are found around the northern and southern parts of the earth. The climate is very hot during the summer and extremely cold during the winter. Cold deserts get some miniscule rain in the form of snow. Hot deserts, on the other hand, are found around the center of the earth. The climate is typically very hot during the day and cool at night, year round. The rainfall is next to nothing in hot deserts. Also the . . . what? Something approaches . . . no, two things approach."

"Something approaches?" Blaine asked.

"Two things approach," Ganzorig corrected.

Both Blaine and Tracey did another three hundred sixty degree turn to look at the desert around them, and neither of them saw a single thing. What could their local expert be talking about?

"My friend, Solongo, approaches," Buri informed as he looked intently in a particular direction. "And it sounds like she may be hurt."

The twins were even more baffled now. Ganzorig was seeing things, and now hearing things that they were pretty sure weren't there.

"Blaine and Tracey, hop on Odgerel with me! We must get to Solongo!"

"Odgerel?" Blaine asked.

"Odgerel is my horse. She is the swiftest horse in the Mongolian desert."

"But can she carry the weight of all three of us?" Tracey asked.

"Of course," Buri responded. "She can carry the weight of ten full grown men without missing a step."

Ganzorig helped the Sassafrases get on Odgerel, and then he

THE SASSAFRAS SCIENCE ADVENTURES

got on himself, sitting in front of the twins. He immediately gave the horse a little kick, and Odgerel took off like a shooting star in the direction of nothing. Nothing the twins could see or hear anyway. Maybe living in the desert had gotten to Ganzorig Buri. Maybe he was a little on the crazy side and was one to chase the wind. Or, maybe he was right. Maybe growing up in the desert had developed in him heightened senses and awareness. Maybe there really was someone named Solongo out there who needed their help. After all, Ganzorig had found the twins in their lonely spot, had he not?

The three raced through the desert on the bouncing back of Odgerel for only about five minutes when both twins saw her. A lone figure draped in a long dark windblown dress, running toward them with outstretched arms. Ganzorig directed his fast horse right up next to the woman, where he quickly jumped off and caught her as she collapsed into his arms.

"Ganzorig!" she exclaimed in a strange mix of exhaustion and exuberance.

"Solongo," he returned. "What happened to you? Are you all right? Why are you all the way out here?"

"It's the Avargatom," the woman named Solongo answered with a dry throat. "They have come to our village and are threatening to take it from us by force, piece by piece."

"What?" Ganzorig asked in alarm.

"They rode in on their horses the day before yesterday and have been taunting us ever since," Solongo continued. "The biggest one stands at the center of our village and challenges our men to different kinds of competitions, but because of his size, no one has stepped up to challenge him. The Avargatom have already taken all of our stoves and all of our animals. And they will just keep taking more, piece by piece, item by item, because no one will face them."

"What about my brothers?" Ganzorig asked. "Would they not step up to fight for our village?"

Solongo hung her head and shook it.

"I'm afraid not," she said quietly.

The tired but pretty Mongolian woman then pulled back some strands of hair that had been covering the side of her face and revealed a large painful looking bruise.

"Solongo, how did this happen?"

"When I saw that none of the men of our village were going to stand up against the Avargatom. I tried to confront them, but I was met immediately by the back of an Avargatom hand. I am sorry, Ganzorig, I failed."

"Solongo, Solongo, look at me! You did not fail. This matter is not over! I will come back with you, and we will save our village."

A small smile formed on Solongo's face. "I knew this would be your attitude. That is why I wandered out into the desert to find you. I knew you were on your way home from college. The desert is wide, but I knew we would find each other."

Ganzorig wrapped up Solongo in a heartfelt embrace. The Sassafras twins watched the two Mongolian friends hugging each other.

As he looked on, Blaine thought, "Solongo really was out there approaching us, but Ganzorig said there were two things approaching. I wonder what or who the other thing is."

Again, almost like he could hear Blaine's unspoken questions, Ganzorig let go of Solongo, turned to face the horizon, and announced, "Something else approaches."

### *Swirling Sandstorms*

Blaine and Tracey followed their local expert's gaze, but just as before, they could see absolutely nothing in the desert.

"It's a sandstorm, isn't it?" Solongo offered.

Ganzorig nodded. "Yes it is. And it's coming fast. We must find some shelter and find it quickly!"

"What?" Both twins' minds questioned. "How in the world could these two tell there was a sandstorm coming? Nothing at all had changed as far as the Sassafrases could see. And if there really was a sandstorm coming, where in the world were they going to find shelter? There was nothing in any direction except for dust and blue and flatness."

"Sandstorms are strong violent winds that stir up loose sand and sediment, carrying it to another location," Ganzorig said as he helped Solongo get up on Odgerel behind where the twins were still sitting on the horse's back.

"These storms often arrive with little to no warning and usually blow through an area quickly; in five minutes or so. They are caused by a front that moves through an arid desert region. These storms can occur around the globe, and they can be small or extremely large. A sandstorm can spread over hundreds of miles. The dust carried by a sandstorm can cause blinding conditions."

Buri remounted and then kicked Odgerel in the haunches. Again the horse took off as fluidly as ever, like she was running passenger-less through a flowery meadow. The twins continued to look around as they rode bareback between the two Mongolians. They still couldn't see anything out there.

"After a storm has blown through, the devastation can look similar to a tornado," Ganzorig shouted as they rode. "Sandstorms can also pick up spores, bacteria, or pollutants and carry them hundreds of miles away. These storms can be devastating and dangerous, but they can also be beneficial. For instance, sandstorms in the Sahara Desert carry nearly twenty tons of particles all the way to the Amazon River Basin, which provides that rainforest with the essential nutrients and minerals the area needs."

As their local expert concluded the information, the Sassafrases' still couldn't see . . . anything, or wait . . . was that a . . . ?

"Is that it—that wall blowing our way? Is that the sandstorm?" Blaine shouted his question as he pointed.

Ganzorig nodded and kicked Odgerel one more time.

"How could he have possibly known that sandstorm was coming," Blaine thought. "Does he have some kind of mysterious sixth sense?"

The Sassafrases hadn't seen the storm at all at first. It seemed to appear out of nowhere, but now that it was visible, it was huge. They couldn't see either end of the sand wall, and its height was growing bigger and bigger.

Odgerel began to run with an urgency that she hadn't had up to this point. The twins hunkered down between the native Mongolians, and Ganzorig fixed his gaze straight ahead, as if he was racing toward a certain location. Blaine and Tracey couldn't see any kind of shelter, but at this point, they completely trusted Buri and his senses.

The wall of sand behind them grew more imposing, quickly chasing them like a hungry predator. Even with Odgerel's speed, the storm steadily gained on the fleeing four. The Mongolian horse's hooves rhythmically hit the sand in a tenacious beat. Her mane whipped sideways in the wind. Her muscular flanks expanded and flexed as she took her speedy strides. Her head and neck bobbed purposefully with brazen determination. The twins guessed this wasn't Odgerel's first encounter with a sandstorm because the animal's urgency hinted to the fact that she seemed to know what was going on. They just hoped this wouldn't be her last encounter.

The horse raced on with all its might.

Ganzorig guided as he shouted out words of instruction to her in Mongolian.

Solongo stayed staunch and silent.

Blaine gritted his teeth

Tracey hung on for dear life.

And the sandstorm continued to chase with fury.

The wall of blowing sand was towering over them. It hadn't hit them full force yet, but it was now so big it looked like it was already over the top of them. There was still no sign of any kind of shelter anywhere ahead of them. Blaine and Tracey didn't know how they could possibly escape. They were fairly positive this was going to end badly for them. They would be pummeled by this wall and buried under tons of sand.

Solongo suddenly broke from her silence. "Look! Look!" she said, pointing at something in front of them.

Ganzorig nodded resolutely, knowing exactly what his friend was referring to. The Sassafrases were clueless and confused, until something ahead came into view. It looked like a small outcropping of rocks.

Could those rocks really provide the kind of shelter this group of four needed to be protected from this storm of sand? Blaine and Tracey just hoped they had enough time to find out.

They could almost feel airborne sand tickling at their ears as the flying wall bore down upon them. Buri guided his horse from the sand into the small rocky area. With a flying leap, Odgerel's hooves left sand and landed on rock. Then, as if he knew this area like the back of his hand, Ganzorig pulled at his horse's reins and guided her to the biggest rock in the area.

As quick as a whip, Buri jumped off the horse, then he helped Solongo and the twins get down as well. All at once, the twins saw why they had come to this spot. Underneath the large rock, there was a shallow cave that was big enough for all of them to walk down into. It would be tight, but even Odgerel could fit down inside the rocky shelter with them.

The four quickly clambered down into the cave. Ganzorig stepped in last, and as he did, he turned around to pull Odgerel by the reins down into the cave to join them. Just as he did, the roaring storm blasted over them like a freight train, carrying sand and wrath. The light of the sun was blotted out as sand and dust particles skipped over the cave's opening, flying horizontally at ferocious speeds. Enveloped in darkness, Blaine and Tracey felt someone hand each of them some kind of cloth.

"Use this to cover your face," Buri shouted out.

The twins immediately obeyed, frantically wrapping the pieces of cloth around their faces. It helped to keep sand and dust from entering their noses and mouths. The four, plus the horse, hunkered down in the small dark cave, hoping this would not be a long storm.

The absence of sight only seemed to enhance the sense of hearing. The howling winds seemed even louder. The sandstorm almost seemed to be growling as if it were alive. The Sassafrases didn't know how much more of the earsplitting howling they could stand. They began to feel sand piling up around their feet. Were they going to be buried in this cave?

Then, as quickly as it had come, it disappeared. All was still. The sandstorm was gone. The four removed the cloths from their faces. They helped wipe off the sand and dust from each other, and then they emerged from the small shelter.

The twins remembered how Ganzorig said the aftermath of a sandstorm could be just like that of a tornado, but as they looked around the Mongolian desert in front of them, they both thought it looked pretty much like it had before the sandstorm. It was big, flat, sandy, and blue.

One thing that was different, though, is that now they were thirsty. They were probably thirstier now than they had ever been before. There was something about being caught in a sandstorm in the desert that made one's throat scream for dampness. But the twins didn't have anything in their backpacks to drink, and it didn't look like Ganzorig or Solongo had anything drinkable either.

Blaine's mind drifted back to when he and Tracey had landed on the sandy beach in Borneo and their local expert, Trisno Kanang, had brought them sparkling glasses of ice cold lemonade. Boy, did that sound good right now.

Again, as if he was tracking with the boy's mind, Ganzorig Buri addressed the unspoken issue. "Let us all get back on Odgerel and get to a place where we can find water. Then, we will quickly get home and see about these Avargatom."

The four sandstorm survivors mounted their steed. They left the rocky outcropping and once again began galloping through the wide flat desert.

The seconds turned into minutes as they galloped, and every minute felt like an hour to the Sassafrases and their dry thirsty throats. They were sure they hadn't been riding all that long since the outcropping, but time had a way of expanding when a person was parched like this.

They whisked through the blue with the speed of the wind. The horse was keeping her pace. The Mongolians were keeping their silence. And the twins were keeping their eyes peeled for anything drinkable.

Blaine's mind was spinning, and he thought his searching

eyes might be spinning as well, but he was pretty sure that right now he was finally seeing something in the wavy blue heat ahead of them.

Yes—he was sure of it! There was a rippling pool of water set between two palm trees!

Blaine licked his lips. He could already taste cool refreshing liquid pouring down his throat, quenching his thirst. Blaine tried to shout out a word about the place up ahead, but when he opened his mouth, nothing came out.

"Oh, no," the boy thought. "What if I'm the only one that sees the pool? I've got to let the others know! I've got to get some words to come out of my mouth!"

The boy tried to shout again. This time, no words came out, but at least a screeching type of grunt did. The desert pool was out there, beckoning for them to come and drink, and if Blaine couldn't manage to speak, they might miss it.

"Look!" the boy finally managed to blurt out. "Look! A pool . . . a pool of water!"

Ganzorig pulled gently on Odgerel's reins, adjusting her strides ever so slightly, to go in the direction Blaine was pointing toward. Blaine smiled and exhaled in relief. Soon, they would all be drinking gallons of cool water.

Odgerel ran with speed straight toward the blue pool and the palm trees. The spot was getting closer and closer…or was it? Now Blaine was confused. The pool seemed to be getting farther away. At the same time, it was disappearing and re-appearing and then disappearing again.

"Where is the pool you saw, Blaine?" Buri asked.

"It's there! No, there! No, wait, now it's there!"

Ganzorig sat up straight, slowed Odgerel down a little, and gazed out into the Gobi with a knowing look on his face.

"What? What happened?" Blaine asked. "I saw a pool of water. I saw palm trees. And now...they aren't there."

"I should have known," Buri said, shaking his head and slightly chuckling. "It happens to the best of us, Blaine. What you saw was not actually palm trees and a pool of water. What you saw was a mirage."

"What in the world is a mirage?" Blaine asked.

"A mirage is when one sees a pool of water that is not there," Ganzorig Buri answered. "It happens because of the temperature difference between the ground and the air. In the desert, the ground is very hot and the air is cool, which causes light to be refracted in such a way that it looks like a pool of water."

Blaine hung his head in dejection.

"Don't be embarrassed, friend," Buri encouraged. "Like I said, it happens to the best of us."

Blaine sighed. He was a little relieved that he was not alone in mistaking a mirage for water, but that didn't quench his thirst. Were they ever going to find water in this expanse of sandy blueness?

## Chapter 9: Avargatom Challenges

### *Desperate Drought*

Blaine and Tracey weren't sure how much longer they could stand without water. They knew that with the touch of a button, they could zip out of this desert to their uncle's lab, where there was plenty of water. But then they would miss out on learning about droughts and oases, and they would fail to complete a part of the earth science leg of their science-learning adventure, which was something they had never done.

Blaine, who was sitting in front of Tracey, turned and cast a weary, yet determined look his sister's way. Tracey read her brother's wordless look loud and clear. They were Sassafrases. And Sassafrases don't give up. And they weren't about to start now, come water or no water.

The twins turned their eyes from looking at each other to gaze up at their stalwart local expert. They knew they could trust this man and they were sure that Ganzorig had the ability to find water in this desert. Then, almost like he had been waiting for the twins to come to a place of peace and trust, Buri announced, "I see a well."

Flooded with hope, both twins looked forward. Sure enough, ahead of them in the steamy blue desert was the short circular stone wall of a well. Within minutes, the two Sassafrases, the two Mongolians, and the one horse were quenching their thirst with cold refreshing water.

Not long after replenishing themselves at the well, the four rode into Ganzorig and Solongo's village. A bit of their hopefulness slipped away as they dismounted Odgerel and witnessed what was happening.

The village was small and quaint, with only a dozen or so small round structures circled around a small mound of dirt. Each of the structures had a wooden frame that was covered in some kind of off-white cloth. The twins assumed these were homes. Off to the side of the circular dwellings were several lines of wooden fences, which formed several pens where animals could be kept.

The village looked cozy and peaceful, but right now it was obviously not. At the center of the village, in the middle of the little round houses, standing on the mound of dirt, was the biggest and meanest looking man imaginable.

He was shouting out, "You weak and worthless Buri tribe! For two days now, none of you have been courageous enough to stand up and challenge me!"

A pride-filled laugh bellowed from the man's throat.

"We have already taken all your animals and stoves. We did it with ease. It was as simple as flicking a fly off a biscuit. And today, we will take more from you! Because we are the mighty Avargatom. And you, you are the miserable, weak little Buri tribe!"

A terrified shudder reverberated throughout the small crowd of villagers who were there listening to the big man's shouts. Arrogant chuckles could be heard coming from the group of gargantuan Avargatom men who were flanking their evident leader. Behind them, to further emphasize their victories, were scores of tied up animals and a pile of a dozen Mongolian stoves, displayed like trophies.

"I am bored with taking what you own by simple brute force. I want to actually compete with one of you weaklings. So, today, we will do something different," the giant Avargatom man continued. "It's something I call 'The Stacking of the Stalks.' Two of us will run out into the fields and cut down a section of corn stalks. We will tie the stalks into bundles and then carry the bundles back here. We will make four stacks of twenty-five bundles. The first one to complete this task wins. Surely one of you little flies can handle a challenge like that!" The man paused and threw back his head as he let out another round of deep, arrogant laughter.

Before anyone could respond, the giant went on. "What are we playing for, you ask? If one of you Buri are victorious, we will give you back half of your animals. But if I win, we will take all of your food!"

A collective gasp escaped from the Buri villagers, as if they had all just been punched in the gut.

"All our food?" Solongo frantically whispered. "Ganzorig, we are in the middle of a drought. It has been a battle just to grow the little food that we do have. We cannot afford to lose our food to these despicable Avargatom! We have to do something."

Ganzorig nodded gravely. "Yes, Solongo, you are right. Let me talk to my brothers, Dariin and Khulan, to see if they will not stand up against this man and accept his challenge!"

"But, Ganzorig, neither Dariin nor Khulan stood up the first two times the Avargatom leader threatened. What makes you think they will stand up now?"

"It is their duty. It is the right and honorable thing to do. Besides, Dariin is first born, Khulan is second, and I am third. I must ask them to take action before I do anything. Solongo, you know this is our Buri custom."

Solongo nodded. She knew what her friend was saying was true, but wasn't ready to accept it. "Ganzorig, you are so much better at bringing in the corn harvest than both of your brothers. You can beat this giant at this challenge! The stakes couldn't be higher. They already have our animals and our stoves. What will we do without food? Surely we won't survive. We cannot outlast this drought if they take all we have to eat!"

Solongo's passionate pleadings were beginning to become more than whispers. Ganzorig reached out and put a gentle hand on his friend's shoulder and looked deep into her eyes with a confidence that she could not mistrust.

"It's all going to be all right," he reassured.

Ganzorig left the place where Solongo and the twins were standing and walked through the small group of Buri villagers. He stopped next to two men, both of whom were a little bigger and stronger looking than he. The twins could not hear what he was saying, but it was easy to see that he was simultaneously pleading and encouraging them.

After a few minutes, the largest Avargatom issued the challenge again from his mound of dirt. "Will it be more of the same, you measly flies? Will none of you step up and challenge me? Do I strike so much fear into your hearts that you will not even fight to save your food? Oh, how weak you truly are!"

He gave another bellowing laugh. "This is it! I will give you thirty more seconds. Then, if none of you will face me, we will take all your food by force!"

Precious seconds immediately began ticking away. Blaine and Tracey looked toward Ganzorig and his brothers with angst, to see if any of them were going to take up the challenge in time.

Solongo was visibly holding her breath.

The villagers trembled in worry.

The giant at the center of the village looked like he was about to give the final call, when slowly but surely, Ganzorig's biggest brother raised his hand.

"I, Dariin Buri," he said, in a voice that sounded shaky, "accept the challenge."

The challenge-issuing giant looked shocked, but the shocked look was quickly replaced with a cruel smile. He was obviously surprised that one of the Buri had actually stepped up, but he also looked pleased in a twisted sort of way. Surely, he was thinking that now he would now have the opportunity to show off in front of the crowd as he crushed his opponent.

"Okay, then, Dariin Buri," the huge man agreed with a cocky grin on his face. "Let's go stack the stalks!"

Within minutes, both Dariin and the Avargatom giant were standing at the edge of the Buri's cornfields with sickles in hand, ready for the "Stacking of the Stalks" to begin. Dariin was fairly big by Buri standards, but he looked like a child as he stood next to the Avargatom man. Another one of the arrogant men, who was serving as the official for the challenge, began a countdown. When he reached zero, the competition was on.

The giant shot out into the fields like a torpedo and began felling corn stalks with his glimmering sickle. The strong wide sweeps of his arm were bringing down whole rows of corn at once. Dariin, however, hobbled out into the fields with a nervous shakiness that he just couldn't seem to get over. He was definitely making an effort, but it was quickly apparent to everyone watching that he had no chance of winning this challenge. By the time he had his first row chopped down, the giant was already finished with the chopping portion and was well into the tying up of his bundles.

Things weren't looking good for the Buri tribe.

Meanwhile, Ganzorig had made his way back to where Solongo and the Sassafrases were standing. As the challenge raged on, he began sharing about the drought his tribe was experiencing.

"In the simplest of terms, drought is an abnormally long period of time, such as weeks, months, or even years, with little to no precipitation of any kind. During a drought, the conditions are typically hot, dry, and dusty."

The twins looked around. These conditions definitely described the current state of this Buri village.

"A 'drought' can be brought on by many different conditions and can have differing effects," Ganzorig continued. "So scientists got together and created four areas, or types, of drought. The first kind is a meteorological drought. This type of drought is caused by lack of precipitation accompanied by winds and high temperatures. Water sources dry up, ground water is reduced, and the soil becomes as hard as rock and cracks.

**NAME:** Drought
**INFORMATION LEARNED:**
A drought is an abnormally long period of time with little to no precipitation of any kind.

"The second kind is an agricultural drought. This type of drought happens when the soil dries out so much that crops and animals are affected. The third type is a hydrological drought. This type of drought is a reduction of surface or ground water, due to overuse. The final type is socioeconomic drought. This type of drought results in the lack of services, such as drinking water and electricity, due to meteorological or hydrological conditions."

The Avargatom leader was now completely finished tying up his hundred stalks, and he was beginning the last task of stacking the stalks into four piles of twenty-five. Poor Dariin wasn't even close to finishing the task of chopping his section of stalks down.

The faces of the Buri tribe members were becoming more and more downcast.

Ganzorig continued as the only Buri with no apparent worry showing through. "Here, in our village, we have mainly struggled with the first two types of drought. But with hard work, creativity, and the use of tribal wells, which continue to produce water, we have managed to keep our animals alive as well as grow some small crops that produce enough food for everyone. Right now, though, it looks like that small amount of food is as good as gone."

Just as Ganzorig finished his last sentence, the big giant threw his last bundle of stalks on top of the last stack to complete and win the challenge. As he raised his arms in victory, Duriin hung his head and a few of the Buri fell to their knees in defeat. They were about to lose all their food by force.

Blaine and Tracey watched with their own eyes, but they couldn't believe what was happening. The Avargatom went through every home and every field to take every single item of food that the Buri had. The Buri tribe seemed helpless, unwilling, and unable to do anything about it.

The Avargatom now had their animals, their stoves, and all of their food. But evidently that wasn't enough for the nasty men because once again the biggest of them all took his place on the small mound of dirt at the center of the village. Like a great white shark with a taste for blood, he issued yet another challenge.

"You might have thought we were finished! That we would leave you in peace." He bellowed and snarled. "But we are not! There is so much more here in your little village that we want to take! I, the largest of the mighty Avargatom, will give you poor little Buri another chance to prove your worth. I will call out another challenge. Let us see if any of you weak miserable Buri flies will have the strength for this task. This challenge will be an archery competition! You Buri are Mongolians, are you not? So surely you can shoot an arrow from a bow, or are all your arms too much like noodles?"

The giant took a moment to laugh at his own joke, before continuing. "For this challenged, there will be two walls of thirty-three surs. These small targets will be placed seventy-five meters away out in the field. The walls of surs will be exactly five feet wide and eight inches high. The first person to either hit all thirty-three surs or completely knock over the wall wins the challenge."

"If one of you wins," the giant paused and laughed like he had just told another joke, "we will give you back a day's worth of food. But if I win, we will take all of your tools by force!"

An inconsolable gasp escaped the throats of many of the Buri villagers. Others remained silent, as they floated through different stages of shock and disbelief.

Solongo reached out, grasped Ganzorig's hand, and looked at him with sorrowful, tear-filled eyes. This time she didn't say a word.

Ganzorig, however, was not looking at Solongo. He was looking over to where his other brother stood. Like Solongo, he wasn't saying a word, but he didn't need to. It was very clear what he was thinking.

Was the second born Khulan going to step up and face the Avargatom giant in this archery challenge? If Khulan would not, the Sassafrases were sure that Ganzorig would.

As if he could feel the stares on the back of his neck, Khulan gave a quick glance back toward Ganzorig, Solongo, and the twins. Then, he looked forward again and faced the giant. He began visibly shaking, just like his older brother Dariin had.

"Will none of you noodle-armed flies accept my challenge?" the huge man shouted tauntingly. "Is there no Buri with courage? No Buri with honor? In thirty seconds, if no one steps up, your tools will be gone!"

Khulan coughed and flinched and started shaking even

harder, but he did manage to get his hand raised.

"I, Kh-Kh-Kh-Khulan B-B-Buri accept your challenge."

"Okay, Kh-Kh-Khulan," the giant mocked in arrogant laughter. "Let's get you a bow and some arrows and commence this challenge!"

Everyone watched as Khulan followed the big imposing Avargatom over to the edge of a field. Khulan was a stair-step shorter than Dariin, so he looked even smaller when placed up against the scary giant. Both competitors were given bows and a large quiver full of arrows, and the walls of targets were quickly set up, way out in the field.

To Blaine and Tracey, it looked like an impossible task to even hit one target, much less hit or knock down all thirty-three. This kind of archery was apparently a Mongolian tradition, so the twins were hoping Khulan could pull off a miracle upset.

The Avargatom man who was officiating the challenge started his countdown. When zero was called, the archery challenge was underway. To everyone's surprise, Khulan released an arrow from his bow before the giant did. All watched as it soared in a high arch heading directly for the wall of surs. With a smack, Khulan's arrow sunk into one of the targets. A sudden hopeful gasp came from the Buri onlookers.

Immediately after Khulan's direct hit, the Avargatom giant's first arrow hit his target wall with such force that it knocked five surs off in one blow.

Khulan got another arrow airborne, but his second shot missed wide right. The giant, however, made another direct hit with his second shot. This time, he knocked eight surs to the ground. The villager's hope was quickly disappearing.

From that point on, it just kept getting worse for the Buri. Khulan got stiffer and slower. He began missing the target by wider margins, while the giant became swifter and looser. His arm moved quickly back and forth from his quiver to his bow, barely visible.

In the end, Khulan had hit only three surs, while the giant had obliterated his entire wall. Again, the big Avargatom raised his arms in victory, dripping with arrogance.

No more gasps could be heard from the Buri villagers. They watched in dejected helplessness as the clan of Avargatom men once again ransacked their homes, taking for themselves every useful tool the Buri tribe possessed.

### *Oasis in the Trials*

The Avargatom men showed no signs of leaving. In fact, their leader was once more ascending the dirt throne he had claimed in the center of the village.

"What more can they take from us?" Solongo whispered through tears. "We have no possessions left. We will all just shrivel up and die and be buried in Gobi dust."

It hurt the twins just to hear Solongo say that, but they could see why she felt that way. There didn't seem to be any silver lining here. They could imagine no possible way for the Buri tribe to turn the tables in this situation. And to add insult to injury, like he had no beating heart in his chest, the Avargatom giant once again issued an arrogant challenge.

"This is just too easy!" He laughed. "Competing against you Buri is like competing against wind or air. Your courage is invisible. Your bravery is non-existent. So, since you cannot stand, since you are too weak to protect what is yours, I will issue my final challenge. This final game is for everything. The challenge is this: Wrestle me according to Mongolian tradition. No weight classes. No time limits. Just two men going head to head with the sole purpose of getting his opponent to touch his knee, elbow, or upper body to the ground."

"And what do I mean when I say this challenge is for everything? Exactly what I said. If one of you can accomplish the impossible and get me to the ground, we will give you back everything. But if I beat one of you, or if none of you will stand up to face me, then we will take everything from you. We will take possession of your homes, your fields, and all the land that has been owned and worked by the measly Buri tribe for hundreds of years. We will take it from you and banish you to the desert to finish living out your miserable existence among nothing but the dust, sand, and wind."

The biggest laugh yet escaped the big man's mouth. He then clapped his hands together and challenged. "Who is it going to be, you flies? Who is going to step up? C'mon, noodle arms, show me what you got. I wanna see wh—"

"Those aren't high enough stakes, you big ugly behemoth," a strong voice called out, interrupting the Avargatom giant mid-sentence.

Shock filled the big man's face. He began looking around for who would dare interrupt him.

It was Ganzorig, and he wasn't finished.

"What a weak challenge you have just issued, you oversized ape. Really? There is no risk at all in that challenge for you, you rotten smelling Avargatom."

"No risk?" the giant questioned. "Who are you—"

"Shut your mouth. I am not finished," Ganzorig said brazenly, with no hint of shakiness. "Let us up the ante by making the winner's pot for this wrestling match a little more interesting. I suggest that when I throw you to the ground and defeat you, you and your hollow-muscled friends will not only give us back everything you have taken from us, but you will also give us the land that has been in your people's possession for generations. When I beat you, you will forfeit the Avargatom Oasis to us, the strong, courageous, and mighty Buri tribe."

The giant began laughing in disbelief. "When you beat me? What? Who? You want the great Avargatom Oasis?"

"That's what I said, Avarga-snob. And I did not stutter," Ganzorig responded plainly.

"Ha! I accept! If you win, your measly Buri tribe can have the Avargatom Oasis as well as all your possessions back," the giant agreed.

"It's settled," Ganzorig confirmed. "So what are we waiting for? Let us get this match started."

The Buri villagers were still scared, but their fear now had an edge of hope to it. The Avargatom still wore a look of arrogance, but they had added a twinge of insulted anger to it.

Solongo looked worried for her friend, while at the same time proud of her friend. The Sassafras twins just stood there motionless, with their mouths hanging open in disbelief. They had never seen one of their local experts get beat up, but they were pretty sure that was about to change. They were inspired by the steely courage Ganzorig had just displayed, but they didn't see how there was any possibility for him to beat the giant in a wrestling match.

Ganzorig was the shortest of his brothers. He was only a little taller than Blaine and Tracey themselves. How in the world could he beat a man twice his size? They were worried for him. And not just for him, but for the entire Buri tribe.

Ganzorig, however, had zero angst on his face. He wore an expression of confidence—one that said he knew he would be victorious. He put on a traditional Mongolian wrestling uniform, as did the giant. Then, he walked with bold steps straight out to meet the huge Avargatom. The two men stood facing each other in half-crouched attack positions and glared into each other's eyes.

"What's the Avargatom Oasis?" Blaine abruptly asked Tracey.

Tracey looked at her brother like he was crazy.

"I don't know, Blaine," she whispered in annoyance. "I hope to find out, but don't you think now might not be the right time to ask that question? Our local expert is more than likely about to get smashed."

"I know," Blaine responded. "I'm just worried and nervous for him, and I don't really want to watch, so that question just kind of popped out. Hey! Do you think I could use our new taser app and zap the giant?"

Tracey looked like she was about to respond but couldn't find the words, when Solongo interjected. "It is okay, Blaine. Your question about what the Avargatom Oasis is a good one. And, by the way, don't worry about the match. We have a bit of time before it gets serious. Wrestlers always stare each other down for a good long while before either will actually make a move."

The twins looked at the two Mongolians facing off. One was towering over the other. Surely, the giant would pounce first.

"The Avargatom Oasis is the place out in the desert where these ruthless giants live," Solongo went on. "That is all I knew for most of my life, but to my delight, when Ganzorig started studying, he began explaining the science of this wonderful desert to me. An oasis is a place in the desert where a pool of water can be found. The pool is formed when water rises from underneath the ground to form a spring. This can occur naturally or it can be the result of a man-made well.

"Thanks to the water, plants

can grow in the surrounding area. Many of the desert animals gather in these regions to drink, eat, and find shade. Oases can be very small, with only a cluster of small plants, or they can be quite large. The Avargatom Oasis is the biggest natural oasis on this side of the desert. You see, in the Gobi desert, there are several underground rivers that feed springs that, in turn, create natural oases. Also, sprinkled across the old trade routes, like the old Silk Road, there are many man-made oases."

Solongo paused and looked at her friend, still crouching there staring at the giant. "If Ganzorig wins this wrestling match and we gain the rights to the Avargatom Oasis, all our problems with this drought will be gone."

She paused again, and her face fell a bit. "On the other hand, if he loses…"

As the Mongolian woman trailed off, the two wrestlers began to circle each other, still crouched and still staring. The twins could understand why Ganzorig wasn't making a move, but what about the giant? Did he actually fear this bold little Buri man?

Then it happened. The Avargatom giant leapt toward his little foe, but with the speed and tenacity of a cheetah, Ganzorig eluded the huge man, whipping around behind him. Ganzorig wrapped his small muscular arms around the left thigh of the giant, and to everyone's awe and amazement, he actually picked the huge man up off the ground. Now the small crowd of Avargatom onlookers were the ones gasping with worry.

Ganzorig lifted the giant high into the air and then slammed the man hard down to the ground. The giant landed flat on his back with a deafening thud and a plume of dust.

There was a moment of silence—no one believed what they had just seen.

Then, all at once, from the Buri side, there arose a great shout of victory. Ganzorig pumped his fist up into the air and placed his foot on the giant's chest as the victor. The big man's Avargatom companions all turned and fled in fear out into the desert, leaving behind everything they had taken from the Buri.

"Well done, my Buri opponent," the giant congratulated in a strained voice from his back. "You are the champion…and the Avargatom Oasis…is now yours."

He then closed his eyes and slipped into unconsciousness.

He opened his eyes, slowly shaking away the unconsciousness that had briefly held him. His sleep had been restless, but not because of any worry or sorrow. His sleep had been restless because he was excited.

Last night, he had finished repairing the machine. Today was the day. The day he was going to zip to wherever those twins were now and abduct them.

He got up from the basement floor, where he had slept. He walked over to his wall of monitors.

"Perfect," he sneered. "Their SCIDAT data is coming in right now, which means the LINLOC coordinates will be forthcoming."

He turned and looked toward the center of the room where the revitalized Forget-O-Nator was standing, looking like a polished port-o-potty.

"It won't be long now, my dear," he crooned, as if the machine could hear him. "Soon I will have some memories for you to erase."

Turning back toward the monitors, he saw what he needed.

"Hmmm…Pakistan," he mused, with a wicked smile curling up his lips.

THE SASSAFRAS SCIENCE ADVENTURES

"Longitude 69° 25' 17", latitude 31° 20' 38". Okay, Trash-a-fras twins! Ready or not, here I come!"

## Chapter 10: Wolves in Pakistan
### *Ascending through the Atmosphere*

As his sight and strength returned, he was positive he had managed to make the landing spot in Pakistan before the twins. He always tweaked the longitude and latitude coordinates just a bit to ensure he would never land in the exact same spot as the children.

He now found himself sitting on the ground just outside a sheep pen in almost complete darkness. It had been morning at 1108 North Pecan Street, from where he'd come, but it was now nighttime here in Pakistan. He looked around a bit more and saw what he thought was the silhouettes of several huge mountains jutting up into the night sky. He also saw a faint flickering light that looked to be shining from the inside of a faraway small house.

The only signs of movement in this lonesome location were coming from the pen. He could hear the sounds of the sheep rustling around munching softly on what he assumed was hay or grass in the pen. The eyebrowless man stood up and walked over to the wooden fence that was safely barricading the animals and keeping them from wandering off into the potentially dangerous night. He reached out and put his hands up on the fence. To his surprise, it was not hard wood he felt but soft wool. He had touched a coat of sheep's wool that was draped over the fence.

Immediately, a wicked plan jumped into his mind. He could cover up in the wool, disguising himself as one of the sheep. Then, he could crawl into the pen and wait for those twins to land among the animals. They would never suspect he was there.

They would just think he was another one of the many sheep inside the pen. When they weren't expecting it, he would pounce on them, knock them unconscious, zip them back to 1108 North Pecan Street, put them in the Forget-O-Nator, and wipe their memories clean.

"Yes, Yes! It's a perfect plan," he shouted victoriously in his mind.

He put his harness and carabiner away and then he grabbed the large piece of wool off the fence and wrapped it around himself. When he was satisfied that he looked as much like a sheep as humanly possible, he hopped the fence, burrowed in among the softly bleating animals, and began his wait.

It didn't end up being a very long one. Within minutes, he heard the sound of something landing in the pen just a few feet away from him—over closer to the fence. His heart exploded with wicked excitement.

"Okay," he thought. "The twins have landed! Now if I can spot them in this darkness . . ." He strained his eyes, looking for their landing spot. Now would be the best time to knock them out—while they were still in their weakened post zip-line state.

He pushed his way through the fluffy white mass until he reached the spot where the sound had come from. But to his surprise, the twins were not there.

"You're not one of mine," he heard a strong voice say from behind him.

The eyebrowless man turned just in time to see a large dark figure loom into sight, only to quickly disappear. A sudden fear ricocheted through his body—it wasn't the twins who had landed here in the sheep pen. The sound had come from this big dark shadow. Who was it? And, why did he feel so afraid?

"You are a wolf," the second statement resounded from behind him.

The Man with No Eyebrows jerked his head around and what he saw nearly sent him into shock—the shadow was now towering over him.

"I know my sheep, and you're not one of them," the shadow declared. "You are a wolf in sheep's clothing!"

The Man with No Eyebrows tried to get a response to form on his lips, but before that happened, the arms of the huge shadow made a swift swinging move in his direction. A long piece of solid wood hit him squarely across the hairless brow with a thundering smack.

Immediately, he fell backward, unconscious, onto the ground of the sheep pen.

The rip-roaring and zip-soaring experience had just come to an end. Blaine and Tracey were tired, but they were definitely not "adventured out." With excitement, they waited for their strength and sight to return so they could catch their first glimpse of the newest science-learning location—Pakistan. It had been nighttime when they had left Mongolia, and they soon found out it was the same here.

"Tracey! I think we've landed on a big pile of fluffy pillows!" Blaine blurted to his sister.

"They're moving and bleating, Blaine. We didn't land on a pile of pillows. We landed in a pen of sheep," Tracey responded flatly.

"Now I see they're sheep, but they are fluffy, are they not?"

Tracey couldn't argue with that, so she changed the subject. "Our local expert's name is Atif Jilani, right?"

"Right, and we are supposed to get information on the atmosphere, clouds, and types of clouds," Blaine confirmed as he looked at the LINLOC app.

"How do you suppose we are going to find him at night in a pen of sheep?" Tracey asked.

Blaine shrugged. "I don't know. Maybe we should let him find us instead."

**LINLOC** SCIDAT

**LOCATION:** Pakistan
**CONTACT:** Atif Jilani
**LATITUDE:** 31° 20' 38"
**LONGITUDE:** 69° 25' 17"

**INFORMATION NEEDED ON:**
Atmosphere, Clouds, Types of Clouds

Now he changed the subject back. "Fluffy, Tracey! They are fluffy! And if my calculations are correct, we have been zipping through earth science for five days now, but we have only gotten two nights of sleep. I wanna find our next local expert; I do, but these sheep are fluffy, and cozy, and warm. We've got to get some sleep, don't you think?"

Tracey looked around at all the gently bleating sheep. She had no idea why her brother was so excited about the prospect of sleeping in a sheep pen. She did not share his enthusiasm, but she figured he was probably right. They did need some rest.

"But, Blaine," Tracey muttered. "What if I can't fall asleep?"

"Then just try counting sheep," Blaine shot back, with a playful smile.

It turned out the Sassafras boy had been right—the twelve-year-olds were tired and they definitely needed some rest. Mere moments after shutting their eyes, both Sassafrases were lying fast asleep among the fluffy animals.

It could have been minutes later or hours later. The twins didn't know, but they both sensed someone approaching them. Blaine's and Tracey's eyes both shot open at virtually the same time. The first thing they saw was a bunch of white wool, so they were still in the sheep pen. The second thing they saw was bright sunlight, so the night was gone and a new day had come. The third thing they saw they didn't actually see, rather they felt it in their minds.

They were sure they could feel the Man with No Eyebrows creeping through the sheep toward them in his big shadowy Dark Cape suit. The twins exploded up out of the fluff and jumped to their feet. They were ready to take this eyebrowless man down, once and for all!

Now that they were standing, they could see they were indeed being approached, but it was not the Dark Cape. Instead, it was three children—two boys and one girl. They looked to be about the same age as the Sassafrases. The three appeared to be somewhat startled because of how the twins had suddenly popped up out of the sheep.

Blaine and Tracey could immediately tell that these kids had a deep confidence about themselves. They were all holding wooden staffs and were all dressed in long flowing traditional Pakistani clothing. Both boys had pillbox caps and the girl had a headscarf covering her head.

One of the boys raised his staff and then swung it down powerfully in the direction of the twins' skulls. But before the staff made contact, the other boy stuck his staff out, blocking the first one from hitting the twins. It all happened so quickly that Blaine and Tracey hadn't even had time to flinch.

"Why are you stopping me from punishing these two, Aazmi? They are thieves!"

"We don't know that for sure, Tariq!" the Sassafras protector rebutted.

"Yes we do! There are only ninety-nine sheep here, and these two foreigners are in the pen!"

As the two boys argued about whether Blaine and Tracey were guilty or not, the girl stepped forward and greeted the twins. "Hello. My name is Javeria," she said and then pointed toward her companions, "and this is Aazmi and Tariq. We are shepherds here in the Karkoram region of Pakistan. Well, actually, we aren't officially shepherds yet, but rather shepherds-in-training."

The twins were still reeling from the near-caning they had almost received by Tariq's staff. On the other hand Javeria's kindness was truly disarming.

"We are the Sassafras twins," Blaine said, reciprocating the greeting. "Blaine and Tracey Sassafras. And I guess we are sort of . . . scientists-in-training."

"Tariq here thinks you two have stolen one of the sheep, but you must know that Tariq tends to be a little bit of a hothead and is always quick to accuse others. Aazmi, on the other hand, would stand up for someone even if he knew they were guilty, but he can also be extremely pious. I tend to stand somewhere in the middle, sometimes figuratively and sometimes literally. I always seem to be breaking up their quarrels. All that said, though, the three of us are actually quite good friends," Javeria shared.

"They are wolves, Aazmi! Wolves! How else do you explain us catching them here in the pen?" Tariq shouted.

"I'll admit, it does look a little suspicious," Aazmi agreed, "but there has to be some other explanation."

"There is no other explanation, Aazmi! They are here to steal and hurt the sheep! They are wo—"

"They are not wolves, Tariq," a deep strong voice suddenly interjected into the two boys' conversation.

Both boys immediately stopped talking and turned their attention toward the voice. Javeria and the twins did the same. A very tall, dark-skinned, strongly built Pakistani man now stood at the sheep pen's gate. He was holding a large wooden staff and was dressed in long dark flowing clothes that looked similar to the Pakistani children's clothing.

"They are not wolves," he affirmed, with a voice that demanded respect and attention. "There is a sheep missing, but these two newcomers did not steal it."

The big man pointed toward the ground near where he was standing, and then he continued a line of pointing that ended somewhere up in the nearby mountains.

"Our lone missing sheep has made tracks," he continued. "It has wandered off by itself in yonder direction, and today the five of you are going to venture into the mountains with me to find this lost sheep."

Tariq looked confused. "You would have us leave behind ninety-nine sheep just to go out and look for one?"

The big man looked at Tariq with the fiercest yet kindest eyes the twins had ever seen. "Yes," he answered resolutely as he turned to follow the path up into the mountains.

The narrow foot path they were taking while following the

tracks of the lost sheep continued to lead higher and higher into the mountains. The big strong dark man led the way. He was followed closely by Tariq, Aazmi, and Javeria, and then Tracey and Blaine brought up the rear. As they got higher, the twins started to feel their lungs burn. They had to work harder and harder for every breath. The view from the trail, however, was worth all those labored gasps.

The mountains reached out as far as the eye could see. The snow-capped peaks were glimmering in the sunlight. The way these mountains towered up into the atmosphere almost made the Sassafrases feel they could reach out and actually physically touch the sky. It was probably the most beautiful view the twins had ever seen.

"The atmosphere is a blanket of gas that surrounds and protects the planet," the big man was saying as he led everyone. "It contains the air we breathe, protects us from being hit by space rocks, and traps heat from the sun, which helps to keep the earth warm."

"Does he always do this?" Tracey asked Javeria as she hiked behind her.

Javeria nodded her head yes. "He sure does. He is training us to be shepherds first and foremost, but he doesn't just want us to be simple shepherds. He wants us to be intelligent shepherds. He wants us to know how to care for our sheep and how to care for our world. He is training us in the ways of animal husbandry and also about the science of everything around us. It should make the hike more enjoyable for you two, since you are scientists-in-training!"

From his spot behind Tracey, Blaine smiled before asking, "Does his name happen to be Atif Jiloni?"

Javeria laughed upon hearing Blaine's question. "Yes, I think that is his original name, but now everyone just refers to him as 'The Shepherd.'"

"The Shepherd?" Blaine asked. "But isn't that a little confusing? Because there are a lot of shepherds in this area of Pakistan, right?"

"Right, but he is not just a shepherd. He is 'The Shepherd,'" Javeria answered. "For longer than any of us have been alive, he has been taking fatherless children under his wing and training them how to make a living by caring for sheep."

The twins nodded in understanding and in a growing respect for their new local expert.

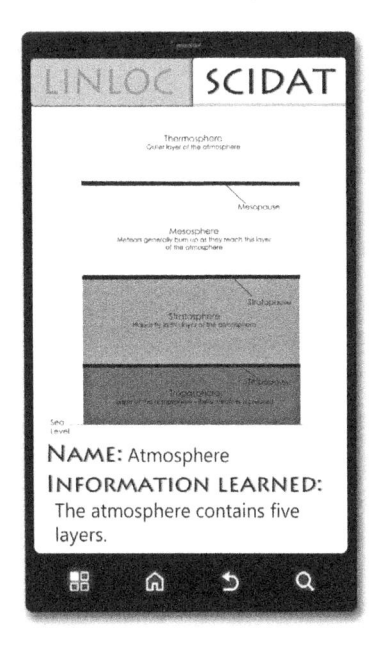

"The atmosphere contains five layers," the Shepherd continued from his spot at the front of the line. "They are the troposphere, stratosphere, mesosphere, thermosphere, and exosphere. The troposphere is the layer of the atmosphere closest to the earth. This layer is where all the weather occurs. Next is the stratosphere, where airplanes typically fly because the air is very still in this layer. The ozone layer is also found in the stratosphere. This layer of ozone gas helps to absorb the harmful ultraviolet rays of the sun. It is susceptible to damage from certain chemicals that are used by humans in things such as spray cans and refrigerators."

The vertical mountain trail flattened out a bit as it led through a field of brilliant wildflowers, giving the hikers a little relief. The trek was still taking the Sassafrases' breath away, but so was the scenery.

"The next layer of the atmosphere is the mesosphere. This

layer is very cool because there are no clouds or ozone to absorb energy from the sun. Many of the rocks from space burn up while passing through this layer," the Shepherd informed them, as he strode through the colors of the flower field. "The thermosphere is the layer above the mesosphere. It is very hot due to the presence of atomic oxygen that absorbs energy from the sun. The aurora borealis and australis occur in this layer."

Blaine and Tracey both smiled upon hearing this last sentence. They thought back to that cold but beautiful night in Patagonia, staring up into the sky and seeing the mysterious dancing southern lights. They were both enjoying how all the science they were learning seemed to be building on itself.

"Lastly, there is the exosphere," the Shepherd said. "This is the farthest layer of atmosphere from the earth. Additionally, you all should know that our sky looks blue in the daytime because of the way sunlight filters through all these layers."

### *Capturing Clouds*

He tended to be a bit of a hothead. He knew that. But surely he was justified in being more than a little angry right now. Once again, he had thought he had the perfect plan to best the twins, but once again his plan had failed. This time the failure didn't come from poor planning or thoughtless missteps. Failure came because he got knocked unconscious by a huge shadow of a man.

The last thing he remembered was getting hit in the face by that stick. He woke up sometime later on the ground outside the pen. He didn't know how he had gotten there, and he didn't want to find out.

As quickly as he could, he got his harness back on, calibrated his carabiner, and zipped back to 1108 North Pecan Street. Now he was sitting in his basement, steaming mad. The Forget-O-Nator sat right in front of him with its door open, like it was hungry for him to throw those twins inside and erase their memories. But he had failed in abducting them.

    Yes, he was mad, but he would not fall into despair. Quite the opposite. His determination to stop the Sassafrases was perpetually growing stronger and stronger.

    The group of six now came upon a pristine alpine lake. The perfectly pure water of the placid lake acted like a mirror, displaying the reflected beauty of the flowers, mountains, and blue sky in an enchanting way. The twins found that if they looked at the lake's surface along with the background it was duplicating too long, they couldn't tell which way was up and which way was down. Nor could they see which image was the real one.

    The trail wrapped around the lake, and then it started going up again. As they climbed up through the last traces of green, the narrow winding trail brought them up above the tree line.

    "At high altitudes, it is typically much cooler because there is less air to trap the heat," the Shepherd said. "Winds are also stronger at higher elevations, and there is a stronger possibility of heavy rain or snow storms."

    Blaine and Tracey were noticing that it was feeling a lot colder and windier way up here, but right now the sky was clear, with no traces of rain or snow.

    "At fifteen hundred meters, humans will start to feel the effects of the thinner air at high altitudes," the Shepherd declared. "High altitude can cause heart rates to rise and breathing to increase. Altitude can also be hard on plant life, so every alpine mountain area has a tree line, where trees and other large plants will cease to grow because of cold temperatures and less oxygen."

The Shepherd was sharing information while he hiked, never taking a single labored step or breath. The twins, however, were struggling for every step and every breath.

"So that's why this is so hard," Blaine thought. "It's the altitude! I knew it wasn't because I was an out-of-shape weakling."

Tracey wondered how they were surviving way up here at high altitude if there wasn't even enough oxygen for trees to grow.

Nonetheless, even with the altitude battling against them, the twins and the other three children continued following the shepherd. With the absence of green came the presence of gray. The trail was now no longer beaten down grass or packed dirt; it was all gray rock. It became apparent that walking the rocky trail was going to be much more difficult than walking the trail as it had been before.

Blaine and Tracey tried hard to only step on the solid and sure rocks on the path, like the Shepherd was doing in front of them all. Sometimes this meant wide sidesteps and big jumps, but that was better than falling or twisting an ankle.

The group had been going up steadily for hours now, making the twins feel like they had climbed through every level of atmosphere that the Shepherd had shared about. They wondered how one silly sheep could travel this far on its own.

All at once, the Shepherd came to a stop, knelt down, and studied the ground. He saw something there that brought a frown to his face. He stood back up and turned his gaze up the mountain off to the group's right, where there was a noticeable sharp, gnarly crag of rocks.

"What is it, sir?" Aazmi asked the shepherd, alarmed. "Is it a wolf? Did a wolf take our sheep from here?"

The Shepherd did not answer. Instead, he started walking slowly and silently up toward the crag.

"Maybe it's not a wolf," Tariq whispered loudly. "Maybe it's a bear or a snow leopard!"

"It really doesn't matter," Javeria interjected. "The Shepherd has faced all the animals you mentioned, and worse. And every time, he was the victor."

"Yes! That is true, Javeria," Aazmi agreed. "So we have nothing to be afraid of!"

"Oh, there is plenty to be afraid of," argued Tariq. "Look how the Shepherd is holding his staff like he's ready to swing it. That means there is something bad up in those rocks."

"You do what you want, Tariq. I am following," Aazmi responded and then immediately started scrambling up behind the Shepherd.

Javeria followed behind Aazmi. The twins looked at each other and instantly agreed, non-verbally, to follow as well. This left Tariq alone, but he didn't stay that way for long. He very quickly decided that heading up with the group was better than standing solo.

The Shepherd effortlessly reached the crag and then quickly disappeared behind a large sharp rock, brandishing his staff as he went. The five children took a much longer time to get up to the spot, not because they weren't trying, but because they weren't as skilled at traversing rock as the Shepherd was. Plus, each had a tinge of fear in their hearts regarding what they might find around the corner.

Aazmi reached the sharp rock first. He paused at the spot long enough for the other four to join him there. Aazmi slowly peeked around the corner and then, all at once, jerked his head back with wide eyes.

"What is it? What did you see?" Blaine asked with excruciating curiosity. "Is the Shepherd fighting a bear? A lion? A Sasquatch?"

"I didn't see the Shepherd," Aazmi responded.

"What?" Tariq shot back. "You didn't see him? What happened to him? What did you see?"

"I saw a cave," Aazmi said.

"A cave?" all four asked in unison.

Aazmi nodded his head.

Neither one of the Sassafrases had a staff, but they had both already developed a deep trust in the Shepherd. So, with a deep breath and a big step, they walked around the big sharp rock and approached the cave. They were followed closely by the three shepherds-in-training.

The mouth of the cave was large, and it was dark. The twins couldn't see more than a few feet into it. There was really no other place to go in this craggy area other than into the cave. So, they assumed that it must be where the Shepherd had gone. Blaine and Tracey drew closer to the cave's entrance cautiously. Their eyes peered into the darkness. Their ears strained for any sound.

Blaine was first to step into the cave, and just as he did, a big strong hand reached out of the darkness toward him. It was the Shepherd's hand, and he was beckoning the children to come into the cave. They all heeded the Shepherd and made their way, a little fearfully, into the cave.

It took at least half a minute for their eyes to adjust to the darkness, but when they did, they could see the cave was as large as its mouth but not very deep. It was basically one big room with only one entrance. There were no wolves, bears, or carnivorous cats in this cave. And there was no sheep either.

"It was here," the Shepherd told the children matter-of-factly. He pointed with his staff to several spots throughout the cave that showed signs of the missing sheep.

"But where is it now?" Tariq asked. "That dumb sheep wandered all the way up here to this cave. We successfully tracked him, and now you're saying we have lost the sheep again?"

The Shepherd put a kind but firm hand on Tariq's shoulder. "All of you, like sheep, have gone astray," the big man said to his shepherds-in-training.

He let the statement sink in for a second before he addressed the situation at hand. "The missing sheep wandered out of the pen on its own. We followed its tracks up to the flower field, which is where I found the clues to something disturbing. There were signs of a slight struggle, making it clear that something took our sheep there in the field."

This statement received gasps and disbelief from the five children.

The Shepherd continued. "I have heard your speculations about which kind of wild animal you thought might have had a hand in ravaging our lost one, but I can now assure you that it was not a wild beast that abducted our sheep. It was a human."

"A human?" Javeria asked. "Who would do such a thing?"

"This should not surprise you, dear one," the Shepherd said to the girl. "There are humans who will parade around as wolves in sheep's clothing, acting like they are full of charity. But in all actuality, they are only full of malice. For example, last night I found a man in our sheep pen who was literally disguising himself as one of our sheep. My staff and I taught him a lesson, but he managed to escape into the night. When I saw that our sheep had been abducted in the flower field, my first thought was of this masquerading sheep-man, but now that we've followed the tracks up to this cave, I am certain that he is not the one."

"How can you be so sure?" Tariq asked. "How do you know it wasn't the wolf-sheep-man?"

"Because this is the cave of the Raider."

Tariq, Aazmi, and Javeria loudly gasped at the naming of the perpetrator. Blaine and Tracey, however, were clueless as to what the name meant.

"Who is the Raider?" Tracey asked Javeria in a hushed voice.

"His real name is Qaiser Qazi," the shepherd girl replied in a hushed tone that matched Tracey's. "People call him the Raider because he is infamous for his stealthy thievery. He is everything the Shepherd is not. He is a thief, a liar, and a cheat who makes his way by dishonest gain."

Both twins recoiled a bit. This Qaizer Qazi didn't sound like a very good guy.

"This isn't the Raider's only cave," the Shepherd continued. "He has many throughout the region. He uses these hideouts when he needs to hide things he has stolen. But he must have known we were coming because he left well before we got here."

The Shepherd paused and looked around the cave. "Our lost sheep is still alive," he mused. "But it may not be for much longer. My guess is that the Raider is headed down to the city with it, where he intends to sell it in the market . . . or worse."

"How could the Raider be so bold?" asked Aazmi. "Does he not know you are a good man? Does he not know how you care for your sheep? Why would he steal from you?"

"Because that's what the Raider does." Tariq answered instead of The Shepherd. "That's why he's called 'The Raider.' Everything he does is despicable. He raids, cheats, and steals, and he never has to pay for his wrongdoing because everyone in the city is afraid of him."

"Doesn't he know the penalty for stealing a sheep is twenty lashes? Doesn't that make him afraid?" Aazmi asked, still doing the questioning.

"He doesn't care!" Tariq answered, still doing the answering. "He doesn't have any fear."

Instead of addressing the boys or the issue at hand any further, the Shepherd quietly stepped out of the cave and gazed out over the mountains. The five children followed behind him and looked in the same direction. The sky was no longer clear. Clouds were beginning to form in large numbers.

The Shepherd remained silent for a long time, with his fierce eyes staring out with intensity. There was no doubt in the twins' minds that he had a plan and that he was going to continue going after the lost sheep. Finally, he opened his mouth to speak, but he didn't address the sheep, the Raider, or the task. Instead, he shifted his focus to science.

"Clouds are made up of tiny drops of water or ice and dust," he began. "They form when warm air holding water vapor cools down. The way clouds look depends on how much water is in them and how fast they form. If the clouds form slowly, they typically spread out in sheets. If the clouds form quickly, they puff up into heaps. Clouds appear white because the water droplets or ice crystals they contain each cause light to be scattered into its different colored components. The result is that our eyes see all the different colored components, together, as white. However, if clouds get thick enough or full of enough water, not all of the light makes it through, giving them a dark shadowy appearance. The darker the clouds are, the larger the droplets of water they contain."

According to what the Shepherd had just said, the Sassafrases were able to immediately deduct that the clouds they were seeing now were quickly forming into rain clouds. Both twins snapped pictures with their smartphones, while at the same time making mental notes about the information they had just heard. They would enter the data later.

The Shepherd turned his gaze from the sky toward his shepherds-to-be. He could tell by looking at them that even after a grueling high-altitude hike, they were all still totally invested in finding this lost sheep with him. Without saying a word, he set his feet to the trail and started back down the mountain. Then, just like he knew they would, all five determined children followed him.

The trail down to the city was a different trail than the one they had come up. It was also more technical. Now they weren't just hopping and side-stepping rocks. They were actually climbing up and down them. The Shepherd led with skill and agility, always looking at the ground for tracks that only he seemed to be able to see. Then, right behind him, there seemed to be an unspoken competition going on between Aazmi and Tariq regarding who could rock climb the fastest. The Sassafrases were going about the same speed as Javeria, as they made up the trio that brought up the rear.

Lower and lower in altitude they went as they crawled up and down the rocks like mountain goats. They quickly got back down to the tree line, and as they did, everyone saw they were approaching a cliff. Everyone, that is, except Blaine. His attention had been hijacked by a cloud that looked to him like a big stack of pancakes.

"Evidently I'm hungry," Blaine thought to himself. "And, oh my, those pancakes look delicious! I can just taste their fluffy goodness with that sweet maple syrup dripping off the si—" Blaine's daydreaming was cut short because suddenly he no longer felt rock beneath his feet. He felt nothing. He was in mid-air, and he was . . . falling!

## Chapter 11: The Lost One is Found
### *Out of the Alto Clouds*

How far down is it?

How bad is it going to hurt?

Why couldn't I have been paying more attention?

These were just some of the questions racing through Blaine's brain as he plummeted down from the cliff. He closed his eyes, clenched his teeth, and waited for the worst, but suddenly his descent came to an abrupt end.

The boy opened up his eyes and found himself dangling in mid-air. Something had snagged the back collar of his shirt. Blaine managed to turn his head back just a little and he immediately saw what it was that had a hold of him. The Shepherd had stuck out his staff and used the end of it to hook Blaine's shirt collar.

"That was a close one, Blaine." The Shepherd looked into the Sassafras boy's eyes as he pulled the twelve-year-old back to safety. Blaine had never been more grateful to be on solid ground. The other children also looked relieved, especially Tracey, who for a few seconds thought she had lost her twin brother for good.

"Gravity is the force that pulls everything towards the center of the earth. Blaine just experienced this principle when he hiked over the edge of this cliff," the Shepherd informed the children.

Blaine let out a big sigh of relief.

"Gravity is also the force that holds the atmosphere in place. As you get higher in altitude, the effect of gravity is less. So the air gets thinner and thinner as you go up, which means there is less air to breathe. Of course, that also means that when you return to lower altitudes, like we are doing now, gravity gets stronger and there is more air to breathe."

Blaine was pretty embarrassed that he had just run off the cliff. He pointed to the cloud that had almost stolen his life. "I was looking at that cloud right there," he said sheepishly. "I must be really hungry because I was thinking about how much it looks like a big stack of pancakes. I didn't even see the cliff. Thank you . . . thank you for saving my life, sir."

The Shepherd nodded his head.

"There are three main groups of clouds—cirrus, alto, and stratus," he began, turning his attention out into the wide sky. "They are grouped according to the height at which they are found in the atmosphere, plus there are also vertical growth clouds known as cumulus clouds."

All five children looked out into the sky as well, drinking in the beauty as the Shepherd spoke.

"Cirrus clouds are high and wispy. They are found above eighteen thousand feet and are composed of ice that high winds have blown into long thin streams. When you see cirrus clouds, it tells you that the weather will be changing in the next twenty-four hours. There are three different types of clouds in this group. First there are plain cirrus clouds, which are typically white, wispy, streamer-like clouds. They predict fair and pleasant weather. Then, there are cirrostratus clouds, which are also usually white, but they are sheet-like clouds that cover the entire sky. They predict rain within twelve to twenty-four hours. Finally there are cirrocumulus clouds, which typically form in long rows of small rounded puffs high in the sky. They are usually seen in the winter time and they predict fair but cool weather."

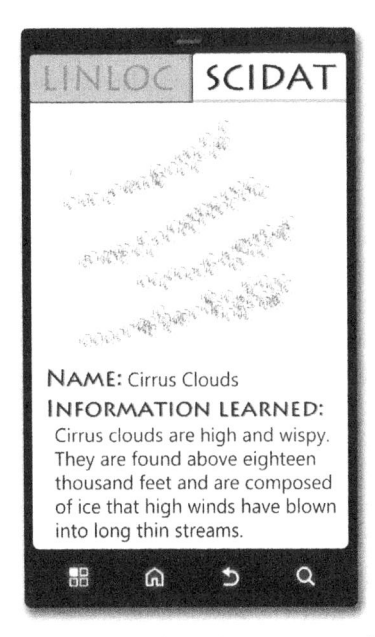

> **NAME:** Cirrus Clouds
> **INFORMATION LEARNED:** Cirrus clouds are high and wispy. They are found above eighteen thousand feet and are composed of ice that high winds have blown into long thin streams.

The Sassafrases turned their attention from the sky to their smartphones. They each opened up the archive application and began perusing through it, looking for all the different kinds of clouds that their local expert was speaking of.

"Alto clouds are mid-level clouds found between six thousand and eighteen thousand feet," the Shepherd continued. "They are typically puffy or flat. There are two main types of clouds in this grouping. Altostratus are usually grayish clouds that cover

> **NAME:** Alto Clouds
> **INFORMATION LEARNED:** Alto clouds are mid-level clouds found between six thousand and eighteen thousand feet. They are typically puffy or flat.

the sky with thinner patches where the sun is slightly visible. They form before a storm with continuous rain or snow. Altocumulus typically are groups of grayish puffy clouds. They form on warm humid mornings and predict coming afternoon thunderstorms."

The Shepherd turned back and looked at the boy who had almost fallen.

"Blaine, your stack of pancakes cloud was a stratocumulus cloud, which is a type of stratus cloud." The Shepherd looked at all the children. "I have shared about these with Javeria, Tariq, and Aazmi, but you and Tracey will have to wait to learn about these groups of clouds. Right now, we must continue down the mountain in search of our lost sheep."

The two scientists-in-training and the three shepherds-in-training all jumped to attention, ready again to tackle the mission at hand. The Shepherd led the children over to the side of the cliff-scape, where there was a trail to guide the group safely down. The rocks and boulders they were climbing down got smaller and smaller in size the further down they went, and the group eventually found themselves on something the Shepherd called a 'rock wash.' It was a wide forty-five degree angled swath of small rocks that were anywhere from pebbled-size down to sand-size.

The Shepherd showed the children how to safely stride down the rock wash by keeping his legs at a wide stance and his body angled with a low center of gravity. The five followed their shepherd down the wash, and all found that this was actually quite a fun way to travel. Blaine and Tracey felt like they were on some kind of rugged escalator as they ran down the wash, the small rocks giving and sliding under their feet with each step. Or maybe it was more like they were flowing down the mountain on a river of small rocks while standing upright. Either way, it was a blast, and they were getting down a lot faster than they were when they had been climbing down.

Eventually, the rock wash came to an end and the trail that

they found at the bottom flattened out a bit. They continued to follow this path down the mountain behind the fearless Shepherd. As they trekked on, the sun sunk lower in the sky and dusk began to set in. Before it was fully night, however, they spotted it; a small city down in the mountain valley.

The Shepherd walked silently down into the candlelit city, through the narrow alleys that wound through the humble mud brick homes, and straight to a particular door. The Sassafras twins didn't know if this was where the lost sheep's tracks had led or if the shepherd had come here for another reason.

The tall dark Pakistani knocked firmly on the door. The party of six waited in the dark alley, but they were met only by silence. The Shepherd knocked again. This time the sounds of someone rustling around behind the door could be heard. It seemed like the person took an abnormally long time to open the door, but when he finally did, he greeted the sheep seekers with a warm smile.

His smile was reciprocated by the Shepherd.

"Naveed!" the Shepherd cried out, as he threw his arms around the man as if he had found a long lost son.

"Dear Shepherd," the man named Naveed responded, hugging the Shepherd in return. "Come in! Come in to my home. All six of you are welcome guests. Please come in!"

The Shepherd walked in through the low doorway first, and then the five children followed. Naveed's home was a very small and humble place. It had no electricity, so the only light was coming from lit candles. Possessions seemed scarce; there was not much more than a few pots, pans, and dishes and a stack of old blankets. The home had one main room and then two small adjoining rooms.

A woman and a small girl were standing with timid smiles in the center of the main room. Naveed walked over, put his arms around them, and proudly introduced the two. "Dear Shepherd, please meet my wife and my daughter."

The Shepherd smiled, walked over, and placed a gentle hand on each female's arm. He looked at them with his fierce yet kind eyes.

"It is very nice to meet you both," he said, in an embracing way that almost made it sound like he already knew them, even though it was obvious this was the first time they had ever met.

"Please, have a seat, everyone. Make yourselves comfortable. Let me get everyone a cup of cool water." Naveed beckoned his six guests to sit down on the floor.

The group sat, and Naveed and his wife and daughter made themselves busy with the task of fetching and serving water. When every guest was holding a cup and Naveed had sat down with his own cup to join them, he explained how he knew the Shepherd.

"I lost both of my parents when I was only eight years old," he shared. "When that happened, the Shepherd was there to take me in. And like he is doing with the five of you now, he taught me how to tend sheep."

Aazmi, Tariq, and Javeria all nodded and smiled. And even though they weren't sure if they were bona fide shepherds-in-training, both Blaine and Tracey nodded and smiled as well.

"I did not end up becoming a shepherd," Naveed continued. "But the Shepherd did teach me a skill set that was very useful, and as you see, I was able to eventually get my own home and my own family."

Naveed put his cup down and changed the subject. "Dear Shepherd, what brings you down from the mountains and into the city at this evening hour?"

Before the Shepherd could speak, the brash Tariq answered for him. "The Raider!"

Naveed looked taken aback by this news. "The Raider?" he

asked.

"Yes, the Raider!" Tariq confirmed. "He has stolen one of our sheep."

Naveed and his family gasped.

Tariq bumbled on. "We tracked the Raider from one of his mountain caves down here to the city. We have to be getting close. He is probably at the market right now, preparing to sell our sheep. Soon, we will find him and take back our lost sheep by force! We will beat that Raider. He will not—!"

The Shepherd reached out and put a calm but firm hand on Tariq's shoulder, bringing the boy's rant to an end. Naveed looked at the group with concern.

"You must not rush out into the night," he warned. "This city can be dangerous enough in the daytime and much more when the nighttime brings its darkness. The six of you must rest here in my home. We have enough blankets for all of you to sleep on. Please stay here tonight. Pause your search. You can resume your hunt for the despicable Raider tomorrow, when the sun is shining again."

Tariq looked like he wanted to respond for the group again, but he didn't as the Shepherd's hand was still on his shoulder. The Shepherd looked at Naveed, his former trainee, and nodded as a simple response to his request upon stay the night.

Blaine and Tracey, along with the other four sheep-seekers, fell asleep quickly. They were all exhausted after their long day of high altitude hiking.

During her deep sleep on the folded blanket, Tracey dreamed that she and her twin brother were back in the sheep pen. Off on a distant mountaintop, they saw the Shepherd victoriously returning with the lost sheep riding safely on his shoulders.

Blaine dreamed he fell off a cliff and landed on a big stack of syrupy pancakes. Then, he bounced from that stack to another stack and another until he finally came to a rest in a pool of butter and syrup.

Both twins woke up happy and refreshed, but also determined. Like the rest of the group, they were determined to find the lost sheep and take down this Qaiser Qazi character in the process, if need be.

The shepherds thanked Naveed, his wife, and his daughter for their hospitality. Then they hopped back on the trail of the missing sheep, making their way through the narrow winding alleys of the residential area and down into the heart of a busy market. There were vendors selling all kinds of items, but it seemed to mainly be a meat and produce market. The twins shuddered as they walked by stalls with raw meat hanging on hooks. They hoped this would not be the fate of the poor lost sheep.

The market was cramped, and the vendors were loud. The further they went, the louder and more cramped it seemed to get. The sights, sounds, and smells of this jam-packed market oozed the feeling that it would be a fitting place for someone known as 'The Raider' to do business. The twins suspected shady deals happened here all the time.

As the group trudged on, vendors began grabbing at the children's clothes, trying to get them to look at and buy their goods. The Shepherd, however, was left alone by all the salespeople as he led the way, like a sharp iron wedge driving its way through thick wood. His very presence seemed to demand respect from everyone he encountered.

Deeper into the sweaty market they went until, all at once, the strangest thing happened. The crowd of buyers and sellers grew quiet and all stepped back to create a wide and open path for the six.

It was a path that dead-ended at the feet of a huge dark brick wall of a man. He was covered in midnight black cloth from his neck to his toes. Strapped around his weathered bald head was the band of a black eye patch, which covered his right eye. He was clutching a man about half his size by the collar and holding him up in the air. As the crowd split, he turned his attention from the little man he was holding to the Shepherd and the five children that were standing at the other end of the clearing. He glared at the Shepherd with his good eye, as he bared a pompous grin.

"Atif," he growled, with a deep low voice, acknowledging the Shepherd's presence.

"Qaiser," the Shepherd responded.

"The Raider," both twins thought. "And he somehow knows the Shepherd."

### *Into the Stratus Market*

The Raider briefly turned his attention back to the poor little man that he was still holding up in the air. "You will pay me what you owe me—or else. Do you understand?"

The man fervently nodded, as he was firmly set back on the ground and released from Qaiser's grasp. He quickly scurried off and disappeared into the crowd, whimpering the whole way.

The Raider turned his body to squarely face the Shepherd. This move seemed to send a wave of intimidation through the silent crowd. The Shepherd's face and body language, however, remained confident.

"What brings you down to the marked today, Atif?" The Raider asked the question, almost like a threat.

"I am looking for a lost sheep."

"A lost sheep, you say," Qaiser responded sarcastically. "Maybe it's not lost. Maybe it's a runaway. Maybe it didn't want to be cooped up in that pen of yours. Maybe it wanted to be down here in the market, where it could serve a tasty purpose."

At this statement, Tariq bounded up, clenched his fist around his staff, and began to rush toward the Raider, but he was stopped by the strong and wise arm of the Shepherd.

"Don't let Atif stop you, boy," Qaiser teased. "C'mon over here and see what happens."

Tariq huffed and stopped, but he still looked like he wanted to give Qazi a good smack with his staff. The other four children, though, stayed hidden behind the shepherd and revealed tinges of trepidation.

The Raider addressed the Shepherd again. "Why did you come down here to find me, Atif? Are you insinuating that I had something to do with your missing sheep? Do you not know the penalty for thievery is twenty lashes? Everyone here knows I never break the law, and I surely don't want to receive any lashes, now do I?"

Qaiser's speech was caked with smug sarcasm. It made everyone want to smack him, but in view of his intimidating physical presence, they simply stayed where they were.

Not the Shepherd, though. He took a step closer to the cocky man and asked, "Where is my sheep, Qaiser?"

Instead of answering like thunder, the Raider answered like lightning. With dazzling speed, he pulled out a shimmering dagger and streaked straight toward the Shepherd. The Shepherd was more than ready for the attack. He threw one end of his staff up and knocked the blade cleanly out of Qaiser's hand. Then, he twirled his staff around, using the other end to smack the black-clothed man hard on the cheekbone, right underneath his one good eye. The Raider fell to his knees and grabbed his face. The crowd gasped, stunned.

The Shepherd waited for the Raider to look up at him, and then he asked again, "Where is my sheep, Qaiser?"

Qaiser scowled and remained silent for a few long seconds. Then, finally he answered, "I didn't abduct your sheep, Atif, but I know who did."

"Why should I believe a single word that comes out of your mouth?"

"Why? Because I am a truth teller of course," the Raider said smugly. "Or maybe I'm not telling the truth. Maybe I am headed home right now to enjoy some mint jelly and lamb chops! I hear that stolen meat is sweet to the palate."

Tariq had had enough. He swung his staff at the Raider, but before he hit his target, Qaiser Qazi snatched the piece of wood out of his hand and used it to take a wide swing at the crowd, pushing all the onlookers back a little. The Raider then jumped up and dashed off behind a stall, knocking a pile of pomegranates to the ground in an attempt to block the Shepherd away from him. He swung Tariq's staff at anyone and everyone as he raced off into the crowded market.

There was still some fear lurking in the hearts of the Shepherd's children, but not enough to stop them from following the Shepherd and giving chase after the fleeing Raider. Qaiser Qazi was fast, but not so fast that the six couldn't keep him within eyesight. They were able to stay on his trail as he raced through the crowded marketplace.

The Raider continued to knock over all kinds of obstructions as he ran, in an attempt to slow down his pursuers. However, they successfully jumped over, side-stepped, and eluded everything that he threw their way. The most difficult thing for the chasers to handle was the crowd. They almost seemed to be trying to help the Raider escape as they crowded the market's aisles. The six maneuvered this obstacle as well, and successfully stayed on the heels of Qazi.

Before they were able to catch him, however, the Raider made his way out of the market and into what looked like a rather large stadium. As they raced through the entryway after him, they all saw that the arena was just as crowded as the market had been, probably because it was currently host to a polo match.

The Sassafras twins had only seen a polo match one time in their entire lives, and that had been on TV. It had been England vs. France, if they remembered right. All the players had been dressed to the hilt. The horses were perfectly groomed and trotted around, obeying their riders' every command. Mallets were gently swung, referees officiated with confidence. If a pin were dropped in the quiet crowd, it would have been heard.

Such was not the case with this polo match here in the Karakoram region of Pakistan. These polo players were unshaven and wore loose-fitting clothing that flapped in the wind. The horses were wild-eyed and tore around the field with streaking manes, snorting loudly, and kicking up dirt with their hooves. It was a wonder to the twins that any of their riders managed to stay in the saddles. Not one of the mallets was being swung gently. Rather, the players were treating them much more like swords in a swashbuckling pirate fight, swinging them around wildly even when they were nowhere near the ball.

There looked to be a couple of referees, but they just seemed to be trying to stay out of the way. The crowd of polo fans was absolutely raucous. It was obvious that they were passionate about the game and viewed the players and their horses like rock stars. The volume that they were emitting was deafening.

Qaiser Qazi raced through the fans and then did the unexpected as he hopped a barricade and ran out onto the field right in the middle of the two teams and their horses. Evidently, the Shepherd really did have him running scared.

The group of six made their way through the crazy crowd

and up to the barricade and then stopped.

"What in the world is the Raider doing?" Tracey asked herself. "Doesn't he know he is probably going to get smashed out there on the polo field?"

Blaine took the momentary pause in the chase to ask himself, "Do you think you could be a Pakistani polo star?" A smile crept across his face as he answered himself, "Yes, I think you could be!"

Qaiser tried to run around the polo playing horse riders, but instead he only accomplished becoming hopelessly corralled by them. The black-clothed Raider got pin-balled around by several of the horses, and then he got hit squarely in the sternum by a mallet. This stopped him and dropped him to his knees in pain, but even this did not stop the polo match.

The players and horses continued to swirl around the fallen Raider like he wasn't even there. All at once, a horse raised up on its hind legs and stood over Qazi. It looked like it would trounce him into oblivion, but before anything too violent happened, a loud strong voice echoed over the arena on a microphone.

"Halt!" it said clearly. "Halt the game!"

Immediately, all the horses and riders stopped. All mallets ceased swinging. Even the rowdy crowd hushed to complete silence. Every eye in the place then turned and looked at a robust man sitting on an elevated platform in the center of the crowd. He was dressed like royalty, and he obviously demanded everyone's respect and attention.

"What is the meaning of this?" he said, with an edge of vexation. "Why did you hop the fence and run in to interrupt our polo match?"

Everyone now knew he was addressing the Raider. The polo players pulled their horses back, leaving Qaiser Qazi kneeling all by himself in the middle of the arena with every stare directed his way.

The Raider did not answer. Instead, he just glared up at the man on the platform.

"Who is the man with the microphone" Tracey asked Javeria in an extremely quiet whisper.

The shepherd girl answered in kind. "That is the Magistrate. He governs over all in this city. His word is law."

Tracey nodded in understanding.

"I ask again: what is the meaning of this? Give me your answer!" the Magistrate boomed forcefully.

The Raider lifted his hand and pointed at the Shepherd.

"It's this man's fault," he shouted. "He attacked me in the market and then chased me into the arena and onto the field!"

All eyes now shifted their focus to the Shepherd. The Magistrate looked like he was about to ask another question, and the Shepherd looked like he may have been forming a response. But before either could speak, Tariq pointed out toward the field and yelled out an accusation.

"This man stole one of our sheep!"

The crowd gasped and redirected their attention to the Raider.

The Magistrate held the microphone very close to his mouth and spoke in a low voice. "Do you not know the penalty for thievery is twenty lashes?"

Still glaring, the Raider answered, "Yes, it surely is, but I did not steal their sheep."

Tariq snorted like one of the polo horses, completely unconvinced.

"No, I did not steal their sheep," Qaiser answered again. "But I know who did, and that guilty person is sitting in this crowd right now!"

Another gasp escaped from the crowd. The Raider lifted his

pointing hand again, and this time directed his finger to a spot behind where the Shepherd and his five followers were standing. Every eye followed the line of pointing directly to one individual—Naveed.

"Wait! What was Naveed doing here?" Blaine and Tracey frantically questioned internally. "It was a preposterous idea—that the hospitable man could be the thief! He was one of the Shepherd's beloved trainees. Why in the world would he steal a sheep from the man who had taken him in and treated him like family? It made no sense."

The people sitting around Naveed scooted themselves away from the man exposing him all the more. Tariq, Aazmi, Javeria, Blaine, and Tracey were all ready to discredit Qaiser Qazi's accusation, but the look they all saw on Naveed's face was making them hold their tongues. The poor man looked completely and utterly guilty.

The Sassafrases' internal questions continued. "Does he look so guilty because hundreds of people are staring at him? Or did Naveed actually steal the sheep?"

The children noticed the Shepherd's eyes were beginning to well up with tears. Not tears of weakness but tears of a strong man that had been betrayed.

"Is it true?" the Shepherd asked his former protégé. "Did you steal one of my sheep?"

Naveed's face slowly got whiter and whiter. He opened his mouth to form an answer, but he barely had enough courage to push it forward from the back of his throat.

"Yes, Shepherd, I did."

The five children gasped along with the crowd. The Raider's face again contained a smug smile. The Magistrate remained silent. The Shepherd stood completely still, leaning on his staff.

"Why?" he asked his former student.

Naveed put his head down in shame, before answering, "I am working with the Raider. He is teaching me how to steal."

"None of that can be proved!" Qaiser shouted from his spot on the field.

"It's true," Naveed said, with his head still hanging. "I am poor, and the Raider told me I could change my fortune by becoming a thief. Your sheep is the first thing I have ever stolen. It was I who did it. I am the one who abducted your lost sheep."

The twins couldn't believe it. Their minds were racing with even more questions, "How could Naveed go from being trained by someone as respectable as the Shepherd to being taught by someone as despicable as the Raider? Why did he not just use the skills the Shepherd had taught him to earn an honest living? Additionally, how had he managed to host them so well last night while being the sheep thief?"

"Your sheep is still alive, dear Shepherd," Naveed continued. "I have it locked in a crate, and I was preparing to sell him in the market. I thought I had wiped away any trace that I was ever near a sheep, but you followed the traces directly to me, didn't you? I knew your tracking skills were good, but I didn't know they were this good. Last night, when you stayed in my home, you knew it was me. You knew I, your former son, was the one who took your lost sheep, didn't you?"

Tears were brimming in the Shepherd's eyes. "I suspected," he answered simply.

Always quick to jump in, Tariq interjected. "But it's the Raider's fault," he shouted. "He made you do it!"

Both Naveed and Qaiser looked like they were about to respond, but it was the Magistrate's voice that sounded out loudly. "The Raider's involvement is inconsequential. It was Naveed that committed the crime of thievery, so it is Naveed that shall receive the twenty lashes!"

Now the tears began running freely down the Shepherd's face. The fans in the arena began either cheering or booing—it was a difficult to decipher which it was. The Raider took this as his opportunity to run off the polo field. The nervous energy in the place began to reach a fevered pitch. Then the Shepherd spoke out suddenly, bringing everything to a halt.

"Sir Magistrate," he addressed the governor. "Is it not true that an innocent party can voluntarily take the lashes for the guilty party?"

The Magistrate's face wrinkled curiously. "Yes, that is legal, but . . ."

"Then I volunteer to take Naveed's twenty lashes."

Silence again fell across the area.

The Magistrate looked down from his elevated platform into the face of the Shepherd. "Are you sure you want to volunteer for this?"

The Shepherd nodded resolutely.

"Very well, then," the Magistrate consented. "The punishment for thievery is to be carried out immediately."

"What?" Blaine and Tracey's minds were screaming. "This is all wrong! The Raider should receive the lashes, not the Shepherd!!"

Almost as if they had been waiting for the occasion, four muscle-bound guards with whips in hand walked into sight and came toward the spot where the Shepherd was standing. They collectively grabbed him and roughly led him to the center of the polo field where the Raider had just been. They ripped off the Shepherd's upper cloak, revealing his bare back, and then the lashes began.

The five children turned their heads away, not able to watch. The Raider had disappeared. Naveed's face was now the one streaming with tears. Dark clouds had gathered in the sky, threatening to soak the arena with a sorrowful rain.

Each lash came slowly, with seconds in between, giving time for the expectant dread to rise. Tariq looked up at the sky with sad eyes and spoke more softly than the twins thought he was capable of.

"Those look like stratus clouds," he gently pointed. "I remember the Shepherd teaching us about these one day as we tended the sheep."

Aazmi and Javeria both wiped tears and nodded. Blaine and Tracey nodded as well, knowing Tariq was gracefully trying to shift their focus away from what was happening.

"Stratus clouds form low flat layers under six thousand feet," Tariq kept talking. "They often block out the sunshine, the Shepherd told us. He said that there are three different types of clouds in the stratus group. First he told us about stratocumulus, which are typically rows of low, puffy, gray clouds with some blue sky peeking out in between. Rain rarely forms from stratocumulus.

**NAME:** Stratus Clouds
**INFORMATION LEARNED:** Stratus clouds form low flat layers under six thousand feet. They often block out the sunshine.

"Then, he shared about nimbostratus clouds, which are typically dark gray, wet-looking clouds that bring continuous rain or snow. However, I think what we are seeing now are the regular stratus clouds. These are usually uniform gray clouds that cover the entire sky, blocking the sun out completely. Light mist or rain commonly falls from these clouds." Tariq was still looking up.

Proving the boy correct, raindrops began to fall as the lashes continued. None of the children had been keeping count because they were trying to block the sound out of their minds.

"There is one more kind of cloud the Shepherd taught us about that Blaine and Tracey don't know about yet," Javeria said, sniffling and speaking up.

"Cumulus clouds?" Aazmi asked.

Javeria nodded and then began reciting what the Shepherd had told them about these types of clouds. "Cumulus clouds are high puffy clouds that can grow vertically between layers. The two main types are cumulonimbus and cumulus. Cumulonimbus clouds are typically large puffy towering clouds with gray flat bottoms that can sometimes be anvil-shaped at the top. Cumulonimbus are often referred to as thunderstorm clouds and are associated with heavy rain, snow, hail, and lightning. Regular cumulus clouds are white and puffy, sort of like cotton balls. They are known as fair weather clouds."

**NAME:** Cumulus Clouds
**INFORMATION LEARNED:** Cumulus clouds are high puffy clouds that can grow vertically between layers.

With that, Blaine and Tracey's SCIDAT data was complete for this location, but their hearts were heavy.

The twenty lashes were finished. The Shepherd had somehow remained silent throughout. The raucous polo crow was quiet, as if they knew what they were beholding deserved reverence. It was finished. The penalty had been paid.

The five children walked out onto the field, carefully approached their teacher, and helped the Shepherd get up to his feet. They gently put their arms around him and began slowly making their way out of the arena. Before they exited, the Shepherd looked up toward Naveed with his fierce but kind eyes.

Naveed was sobbing uncontrollably, but he did manage to mouth the words, "I am sorry," in a sincerely repentant way.

The Shepherd's eyes sparkled and his silent smile clearly responded, "You are forgiven, my friend."

## Chapter 12: Back in Alaska
### *A Break in the Water Cycle*

The Sassafras twins were enjoying the light-speed zip-line ride like they usually did, but during this particular ride, they were a little more somber than normal. Everything at their Pakistan location had ended up working out okay. They had gotten all their SCIDAT data and pictures. The lost sheep had been found and returned safely to the other ninety-nine at the Shepherd's home.

But a lingering somberness remained as they zipped across the world. The twins had been deeply affected by the sacrifice they had seen the Shepherd make. It was an event that made them want to cry and yet altogether rejoice at the same time. It was an act of selfless kindness the Sassafrases would never forget.

Their zipping came to a jerking stop. Their carabiners unclipped from the invisible lines. Their bodies slumped down, void of normal capabilities.

As sight and strength slowly returned, the twins were both thinking they knew exactly what was about to happen. First, they would fall through a hole in the ground and go tumbling down a slide into an underground room. Then they would be greeted with screams, hugs, and giggles.

LINLOC had shown Alaska, longitude 67° 3' 58" N, latitude -163° 4' 12". They would be studying fog, the water cycle, the nitrogen cycle, and the phosphorus

**LINLOC**
LOCATION: Alaska
CONTACT: Summer Beach
LATITUDE: -163° 4' 12"
LONGITUDE: 67° 3' 58"
INFORMATION NEEDED ON: Fog, Water Cycle, Nitrogen Cycle, Phosphorous Cycle

cycle. Their local expert was the one, the only, Summer T. Beach.

Sure enough, as the twins stood to their feet, they saw they had indeed landed in that familiar wide field in Alaska. It was big and flat and surrounded in the distance by trees. Now the hole in the ground would open and they would fall onto a spiral slide that would carry them down to Summer's sleek underground science lab where they would be greeted enthusiastically by their favorite local expert.

The twins waited . . . and waited . . . and waited . . . why was it taking so long?

They waited some more. The first two times they had come to Alaska, the ground would have already opened by now, but still nothing happened.

"What do you think is going on, Blaine?" Tracey asked her brother. "The longitude and latitude coordinates were the same as the first two times we came here, right?"

"I think so," Blaine shrugged. "Maybe this ti—ouch! Something hit me! Ouch! Trace, are you? Ouch! Ouch!"

THE SASSAFRAS SCIENCE ADVENTURES

Tracey looked at her brother curiously. Why in the world had he started jumping around and saying ouch?

"Blaine, what is wro—" Tracey started to respond but suddenly stopped her sentence as she felt something hit her in the leg.

"Ouch!" the girl yelped.

Tracey looked down to the spot on her leg and noticed there was now a small blotch of purple there. She looked back at her brother. There were purple spots on him, too. Tracey's mind was full of questions, but suddenly a single one moved quickly to her lips, "Why is there a pack of brown dogs running toward us?"

Blaine answered, "I don't know, but I think we should run!"

The twins took off across the field toward the distant trees in the direction opposite of the approaching dogs. All at once, the Sassafrases' four eyes spotted an individual running at them from the exact spot in the trees they were trying to run to for safety.

As they got closer, they saw the individual was wearing camouflage clothes and dark face paint. The dogs were barking and gaining ground. The person approaching was getting closer and closer. The twins wondered if they should switch courses, but they found themselves continuing to run straight toward the rapid approaching person.

Wait a second. Was this running person holding a gu—?

Louder and closer barking interrupted that question. The twelve-year-olds were certain that some sort of collision was about to happen.

At basically the same time, Blaine and Tracey were met by the face-painted individual and the dogs in a tumbling pile of camouflage, brown fur, and purple blotches. The twins ended up on their backs in the field.

"Hip hip hooray!" they heard unexpectedly. "It's the return

of the Twinky-frasses!"

They knew that voice.

Suddenly, the boy and girl were being licked—not by the owner of that voice, of course, but by the pack of brown dogs.

Blaine and Tracey managed to sit up, but they were immediately knocked down again by a giggling, tackling, hug from the camo-clad character, who just so happened to be their local expert, Summer Beach. Summer pulled the twins to their feet, looked into their faces with a huge smile on hers, clapped emphatically, and then pulled them up in another hug. This time it was not a tackling hug; it was a jumping dance hug.

"Oh me, oh my, oh me, oh my. I am so happy I could cartwheel up into the sky!" she squealed in an elated voice. "I am so glad you two cuties are back in Alaska! And I get to be your local expert for the fourth time—the science learning continues!"

The jumping dance hug ensued for a good while longer, and then Summer finally let the twins out of her grasp, sort of. She was still holding one of each of their hands. Blaine and Tracey were glad to be back here reunited with Summer Beach, but they did have a few obvious questions.

Blaine started with, "Summer, why are you dressed up in camo?"

This was followed up with Tracey's, "What are these beautiful dogs doing here?"

Summer just smiled and pointed at the twins' purple blotches. "I bet you're wondering about these, too, aren't you?"

The twelve-year-olds nodded.

"It's that time of year, here in our neck of the woods, for P.B. and J!"

The Sassafrases looked at Summer and then at each other, curiously.

"It's time for Peanut Butter and Jelly sandwiches?" Tracey asked.

"Which you only eat once a year?" Blaine added.

"No, no, not peanut butter and jelly, you silly-sasses," Summer laughed. "It's time for our annual Paintball Jamboree!"

Blaine and Tracey were still utterly confused.

"Everyone here in our small corner of Alaska," Summer explained, "gets decked out in camouflage, grabs their paintball equipment, and we all get together for this one day to play paintball games as a community. We call it the PB and J – the Paintball Jamboree!"

"P.B. and J?" the twins asked.

"Yep," Summer confirmed. "But this year they accidentally scheduled a moose hunt during the same time, so we have a super low turnout. The two of you getting here has raised the numbers of human players by fifty percent!"

"Human players?" Tracey asked. "You mean there will be non-humans playing?"

Summer nodded, smiling.

The Sassafrases weren't sure if they wanted to play paintball with non-humans, but then again, it could be fun, even though getting hit by a paintball can sting something fierce.

"So these purple blotches we have on us are busted paintballs, right?" Tracey said, putting two and two together.

Summer nodded.

"But who shot us?" Blaine asked.

Summer raised her hand. "I must confess," she giggled. "It was me!"

"Wow! Shot by our own local expert." Blaine shook his head with a smile.

Summer's giggling continued, and then she suddenly started clapping again. "Oh, I am so excited! And guess what. We aren't just going to play paintball. We are going to mix some exciting science into all the fun! It would just be a shame not to! You see, the part of Alaska we are in right now is in the boreal part of the coniferous forest—a.k.a. the taiga. FYI: the name boreal comes from the Greek god of the North Wind who was named Boreas. Hopefully, this isn't TMI."

"The boreal coniferous forest is the world's largest biome. It receives mostly snowfall for its precipitation. It does rain during the summer months, but the ground, rivers, and lakes get a significant amount of their water from snowmelt. The temperature changes drastically from summer to winter. In the summer, temperatures usually range from thirty to seventy degrees Fahrenheit But in the winter, the range is from thirty degrees Fahrenheit all the way down to negative sixty-five! SWIM!"

"S. . . W. . . I. . . M. . .?" the twins wondered. "Did Summer really mean to say SWIM?" They understood everything their local expert was saying about the forest, and they understood the meaning of FYI and TMI, but what in the world did SWIM mean?

Summer saw the look of confusion on her favorite twins' faces.

"I guess you don't SWIM?" she quipped. "It stands for 'See What I Mean?'"

"See what you . . . Oh!" Now the Sassafrases got it. "SWIM . . . See What I Mean."

Now that the children understood, Summer continued. "Because of the crazy range of temperatures, the growing season in the taiga is very short, and many animals migrate in and out. But right now for PB and J, it will be a teaming with life! There will be all kinds of animals out there running around in the forest as we shoot each other with paintballs. But it will also be pretty soggy due to all the snowmelt and intermittent rain."

Summer paused and clapped her hands. "Any-hoo, as you can see, science is happening all around us. We will go down to the lab later and upload some data, but I just wanted to give you two Sass-a-ma-frasses a chance to run around in this amazing forest for a little while. Oh! I almost forgot! Not all the uploading will be happening down in the lab because I am fairly sure I have figured out a way to work this into the game!"

Summer reached into a camouflaged bag she had strapped over her shoulder that the twins hadn't even seen (because it was camouflaged) and pulled out something they both recognized.

"It's the see-through tablet you showed us in Paris!" Blaine exclaimed.

"That's right," Summer responded. "This tablet, along with one more exactly like it, will be used in the capture the flag paintball game we will start in a little while. I will tell you more about that in a second, once everyone gets here."

The twins wondered if the 'everyone' Summer had just mentioned included any of the non-humans she had referred to earlier.

For now, let's take a look at the water cycle as we wait! Smiles found their way to Blaine and Tracey's faces as Summer began reading SCIDAT data from the see-through tablet about the topic of the water cycle. "The amount of water we have on earth is limited. Water regularly changes form in a process we call the water cycle. The process includes evaporation, condensation, precipitation, and collection.

"The first step of the water cycle happens when the sun's heat evaporates water from the world's oceans, lakes, and rivers. Then, the water rises into the air as water vapor. This water vapor condenses to form clouds. As the water vapor cools, it condenses, the clouds become heavy, and the water falls as rain or snow. And then finally, rainwater and snowmelt flow back into the world's oceans, lakes, and rivers to begin the cycle again."

Summer concluded reading and pushed the upload button. Blaine and Tracey immediately felt their smartphones vibrate with the received SCIDAT data.

"I doubt you two can snap a photograph that would work for the water cycle," Summer said. "But I bet you can find an image in the archive app that will represent it well."

The twins nodded, but they weren't the only ones that were excited about all this scientific knowledge.

"Oh, my goodness!" she squeaked. "Have I mentioned how excited I am that you two are here again? Yotimo will be happy to see you again, too—especially you, Tracey. Well, he doesn't really speak . . . or smile, but I'm fairly certain he is happy to see people he knows again."

"Wait . . . Yotimo?" Tracey asked. "He is the Eskimo man who rescued me from the polar bear with his team of sled dogs, right?"

"That's right," Summer confirmed, smiling. "And, speak of the doggies, these are Yotimo's dogs right here!"

Tracey looked at the pack of brown dogs that were now prancing around them happily. A peaceful smile came to the girl's face as she thought back to the time when these sled dogs, along with their musher, had scooped her up out of the snow, had carried her safely along a mountain trail, and had reunited her with Blaine and Summer.

"Where is Yotimo now?" Tracey asked.

Summer pointed to the edge of the forest where the twins had first seen her.

"Actually, here he comes now. Oh, and look! He has Skeeter and Tina with him!"

"Skeeter and Tina?" Blaine asked.

"Sure enough," Summer confirmed. "Skeeter and Tina Romig. They are newlyweds who just moved up here from the lower forty-eight. They will be starting their first semester as elementary school teachers here in a few weeks. You two will like them. They are really nice."

The three approaching individuals quickly got to the spot where Summer, Blaine, and Tracey were standing. The Romigs immediately greeted the twins.

"Wow! It's Blaine and Tracey Sassafras, live and in person! We feel like we already know the two of you because Summer has told us so much about you."

The male Romig, who looked a little like a lumberjack, said, "My name is Skeeter, and this is my lovely wife, Tina." Skeeter paused with a smile and then exclaimed, "It is so cool that you two are learning about the science of our planet!"

"I hope our students will be as eager to learn as the two of you are," Tina added.

Summer looked at the twins and smiled like a proud aunt. "Oh, Skeeter and Tina, you are so right! These cute little twins are just the best! They are learning tons and tons of science and doing such a good job at it!"

Blaine and Tracey both blushed a little as they were somewhat embarrassed by the sudden attention. It was hard to believe that, at the beginning of the summer, they had sincerely hated science, and now they were being commended for their love of it and their knack for learning it.

Yotimo stood there also, but he wasn't dressed in camouflage like the other Alaskans. He was wearing tough and furry clothing that looked to be made of animal skin. He didn't have a paintball gun like the other three, but he did have a large knife secured in a sheath hanging from his belt. Yotimo was not smiling. Rather, his leathery wrinkled face was as solid and unchanging as stone. He didn't look angry. He didn't look happy. He just looked . . . serious.

Blaine briefly remembered meeting the Eskimo man, but he didn't really know him. The twelve-year-old boy reached out and shook the man's solid hand. Tracey, however, did know Yotimo. He had saved her life, and the Sassafras girl needed more than a handshake. Tracey stepped over and gave Yotimo a big hug.

The tall strong man stood completely unflinching with Tracey wrapped around his waist. The girl eventually let go and then looked at the man with friendly and appreciative eyes.

"It's so good to see you again, Yotimo," she smiled.

"Oh! That was so sweet!" Summer said sincerely.

Then, not one to miss out on a good hug, Summer Beach went around and gave everyone in the group a big happy hug. She even hugged the dogs.

### *PB and J with a Side of Fog*

"Okay, now let's get to PB and J!" the scientist said enthusiastically. "Here is how the first game is going to work. The six of us humans plus all these cute little puppy dogs are on one team. The second team is out in the forest. It is made up completely of non-humans—my lab assistant, Ulysses S. Grant the arctic ground squirrel, and all of his robot squirrel buddies!"

Blaine and Tracey shot worried glances at each other. They knew these robot squirrels, and they didn't much care for them. The last time the Sassafrases had visited Alaska, they had been chased all over Summer's underground lab by these metallic vermin. At first, the robots had been playful and friendly, but then for some reason, their eyes had switched from green to red and they had gone haywire. They had started chomping at the twins and ferociously chasing them all over the place.

Supposedly, Ulysses S. Grant had fixed whatever had gone wrong with them, but the whole experience had left such a bad taste in the Sassafrases' mouths that they weren't sure they could trust these little robots again. It was apparent that Summer was not at all worried. In fact, she seemed very excited about the prospect of playing paintball against the small army of robot squirrels.

"Blaine and Tracey, you are never going to believe the modification that Ulysses applied to the robots specifically for the PB Jamboree!" Summer said with bright eyes. "They can still climb up and burrow through almost anything, but he changed the internal specimen holders to carry paintballs! They can now shoot paintballs right out of their mouths! Isn't that cool??"

Another worried look was exchanged by the twins. Summer either didn't see it or maybe she just joyfully disregarded it as she continued on with the parameters for the upcoming game. "Like I said, it is going to be a capture the flag style game except that instead of using flags, we will use two translucent tablets—both of which have valuable SCIDAT data available for uploading!"

The mention of "SCIDAT" suddenly had Blaine and Tracey very interested.

"Our team will hide our tablet somewhere in the field, while the squirrel team will hide the other tablet somewhere in the forest. Our goal is to guard our tablet while finding theirs. And the other team's goal is to find our tablet while guarding theirs. If you get hit by a paintball at any point during the game, you have to put down your gun and sit out until the next game begins. Pretty simple rules, right?"

Everyone nodded as the scientist reached into her camouflaged bag again. This time she pulled out two paintball guns, two camo vests, and two pairs of protective goggles.

"Okay, Sassy Frassies, here's your paintball getup."

Each twin grabbed their respective guns, vests, and goggles. Their anticipation of the game was growing. Summer opened up a canister of brown face paint and dabbed some all over each twin's face. The scientist paused and smiled at her handiwork like she had just painted a Rembrandt. Then, she set the canister down and held up a whistle.

"When Yotimo blows this whistle, the game will officially begin," she announced. "But before we decide where to hide our tablet and start the competition, let's add a bit of scientific motivation to this game! If we can get to the squirrel team's tablet without getting hit by paintballs, we can upload the information from that tablet as well. It is data about fog, if I remember right."

Summer suddenly reverted into a fit of jumping and clapping. "Ahhh! This is all so science-y and so exciting! It's scien-citing!"

Everyone jumped and danced around with the hyper Summer for a little bit, everyone that is except Yotimo. Then, they got down to the business of hiding their tablet.

At first, the twins thought it was a little unfair that they had to hide their tablet in the field while the squirrels got to hide theirs in the forest, but as they strategized, their thinking changed a little. Sure, their tablet may be easier to find in the field, but it also was easier to guard because it would be easier to see the robots approaching. Either way, the game was about to start. Yotimo held the whistle in his mouth with his cheeks puffed out ready to blow.

The whistle sounded out.

The game was on.

What in the world were those twins doing? Were they playing paintball? Summer Beach was as silly and unpredictable as Cecil Sassafras! One could never guess what she may have those twins doing next, all under the banner of science learning.

He hunkered down a little lower in his wooded hiding spot, as Summer and the others spread out with aimed and ready paintball guns. There was also a big strong-looking Eskimo man with sixteen or so dogs, but they were all standing mostly stationary in the field as if they were guarding something.

He scratched his forehead right where the eyebrows would have been for a normal person. This abduction attempt was going to be tricky.

He had missed nabbing these twins in situations that were far easier than this one. How was he supposed to sneak up on them and grab them when there were dogs and paintball guns all over the place? Maybe he should go back and get the Dark Cape suit and try that one more time. Maybe he should wait a little longer . . .

No! No! What was he thinking? He couldn't chicken out again! He had resorted to that tactic too many times already!

He was going to stay and he was going to figure out a way to nab these twins. He was going to zip them back to his basement. He was going to stick them in the Forget-O-Nator, and wipe their memories completely blank.

All the science they had learned would then be gone. Cecil Sassafras would become a crushed failure. His revenge over his nemesis would then be complete. He was going to stay here in Alaska. He was going to figure out something.

He hunkered down a little lower in his wooded hiding spot. Fear of the robot squirrels was creeping into Blaine's heart. How in the world was he supposed to sneak up on and run away from these machines? This wasn't at all like playing paintball against other humans, which Blaine had experienced only once last summer at Camp Zip Fire. Humans were much easier to spot in a forest than tiny robots. Plus, humans couldn't shoot paintballs out of their mouths.

The Sassafras boy took a deep breath and gripped his paintball gun tightly. He would stick to the battle plan his team had mapped out. He would also stick to the hope that this could end up being fun.

She hunkered down a little lower in her wooded hiding spot and let the beating of her heart slow down a bit. Tracey ran through the plan again in her mind. Yotimo and the dogs would stay in the field and guard the tablet. Skeeter and Tina would flank wide to the left. She and Blaine would flank to the right. Summer would go right up the middle, straight into the area of the forest where they thought the squirrel team may be hiding their tablet. They would wait for the squirrels to shoot first so they could pinpoint the location of at least some of the robots. They figured wherever the highest volume of robot squirrels were, was probably also where the tablet would be.

Summer would cause some kind of distraction. Then, either the Romigs or Blaine and Tracey would sweep in for the tablet. It seemed like a fairly good plan to Tracey, but that fact still hadn't wiped away the fear Tracey was feeling about meeting these robot squirrels again.

They hunkered down a little lower in their wooded hiding spot. They knew they were not supposed to be hiding in the exact same spot, but they just couldn't help it. They were newlyweds, after all, and they couldn't stand being apart from each other. They had only been married a few weeks, and they were both so excited about the adventures that lay ahead. Living in rural Alaska and teaching children of low-income families had always been a dream for each of them. Now it was their dream together, and they were living it! They had really enjoyed meeting their new neighbors, Yotimo and Summer Beach. And how about those Sassafras twins!

Suddenly, the Romigs' train of thought was interrupted by a "Pop! Pop! Pop!" sound. Both Skeeter and Tina smiled. Somebody had fired some paintballs. The game was a go!

She hunkered down a little lower in her wooded hiding spot. Either Ulysses or one of the robot squirrels must have seen her because she had just been shot at, but luckily she hadn't been hit.

Summer quietly giggled—this was so much fun! She just loved PB and J! This year was especially fun because Blaine and Tracey were here. Summer was just bubbling over with joy. She was getting to make so many fun memories with Cecil's niece and nephew. She wanted to stand up and start dancing she was so happy. Hey, wait a second; that was actually a great idea! She was supposed to cause a distraction, right?

Oh my goodness. Was he really seeing this? Blaine couldn't help but chuckle. Summer had stood up from her hiding spot and had actually started dancing. Now there were paintballs whizzing all around her. They were being shot from squirrel mouths, but she was swinging and swooping and eluding all the shots, at least for now. Summer Beach really could make anything fun.

The sight she was currently beholding had chased all fear away and had replaced it with laughter. Tracey watched with an entertained smile as her favorite local expert danced around, dodging paintballs. "It's easy to miss a moving target," Tracey thought. "Especially when the target is moving like that."

Skeeter and Tina held each other's hands and quietly laughed. The sight of their new neighbor dancing in camo, dodging paint-filled bullets was a hilarious sight. They were going to love living here. They just knew it.

"What an idiot," he thought to himself. "What was that Summer Beach doing?"

She looked like an absolute fool. Oh, never mind all that. It didn't really matter. What did matter is that he had managed to move just a little closer to where the twins were hiding. They weren't in the same spot, but they were only about twenty yards away from each other.

So far, he was positive his presence had gone unnoticed. If he was going to pull this abduction off, he must remain hidden!

Oh, no! His heart suddenly dropped as he saw a sight that shocked him. There, up in the branches of a tree, holding a tiny yet powerful looking paintball gun, was Ulysses S. Grant! The arctic ground squirrel that had beaten him about a week ago in hand-to-hand combat.

The familiar feeling of fear began to rise up in his throat. He wanted to zip away from here and bother these twins some other time, but then he saw something that brought pause to the rising fear. He saw something dark and gray. There was a thick fog rolling into the forest.

"Just what I need," he thought to himself. "This fog will provide the perfect cover for me to pull off my plan."

She was laughing, jumping, clapping, and dancing! She had even thrown in a few cartwheels to mix things up a little bit. She was amazed that no paintball had yet found its mark on her, but that was a good thing. Hopefully this long distraction was giving one of her teammates enough time to make a charge for the squirrels' tablet.

"What am I doing just sitting here?" Blaine thought. "I have to do more than just watch my local expert dance." The Sassafras boy scanned the woods for the information-filled tablet. He knew it was going to be next to impossible to spot a clear tablet in all of this green and brown, but that was what he was going to have to do in order to win of the game. He had spotted several of the robot squirrels who were still firing at Summer to no avail. Blaine had even spotted Ulysses S. Grant, who was perched up in a tree, but the boy had yet to catch a glimpse of the tablet.

He wondered if any of the squirrels had gotten close to finding his team's tablet in the field. He assumed they hadn't because Yotimo and his band of sled dogs were extremely capable defenders. If they were going to win, now was the time to find the tablet and make a move, but then the boy noticed fog rolling in.

Tracey saw the fog, but she also saw something else. There, at the base of the tree where Ulysses S. Grant sat, was the object they were seeking: the translucent tablet. The Sassafras girl was now only a few yards away from the tree and the tablet, and it was prime time to make a move. The girl also saw the prize was being guarded by at least four robot squirrels. How on earth was she going to get to it?

Finally! Blaine had spotted the tablet. It was at the base of Ulysses's tree. Could he get to it before the fog enveloped the game? Could he grab it without getting pegged by a paintball? Wait a second. Tracey was a lot closer to it than he was. All at once, the Sassafras boy knew what he needed to do.

He steadied his paintball gun and took aim. Five shots rang out in repetition. Splat, splat, splat, splat, splat. Ulysses S. Grant and four guarding robot squirrels were hit. Immediately, like a cheetah pouncing on a wildebeest, Tracey shot up from her spot and raced for the tablet.

With a deafening splat, she was hit. The paintball had smacked her squarely on her left shoulder. She put down her paintball gun, signifying that she was out of the game. But she was not sad. She was happy. Summer smiled because at the precise moment she had been hit by the paintball, she had seen something else happen. Tracey had gotten the squirrels' tablet! The Sassafras girl had won the game for their team!

Ten or fifteen minutes later, all the humans, animals, and robots found themselves standing in the open field as one group. The squirrels had conceded the victory with no qualms. All the eyes of the robot squirrels remained green, which had brought the twins great solace.

Their team had won fair and square, so now it was time to read and upload the SCIDAT data the squirrels' tablet held.

"Fog is a thick cloud of tiny water droplets that are suspended very close to the earth's surface," Summer Beach read aloud, as the actual fog began rolling into the field from the surrounding forest.

"Fog forms in the same way clouds do. The water vapor in the air close to the ground condenses and forms a cloud. So basically, fog is a cloud on the ground. Fog typically only happens when it is very humid, as there has to be a lot of water in the

**NAME:** Fog
**INFORMATION LEARNED:**
Fog is a thick cloud of tiny water droplets that are suspended very close to the earth's surface.

air for fog to form. There are several different types of fog, which are classified by how it forms. There is radiation fog, advection fog, and valley fog. Radiation fog forms as humid air near the ground cools overnight. So this type of fog is usually seen in the morning and 'burns off' as the sun starts shining."

"Advection fog forms when warm moist tropical air moves over a cooler surface. It typically forms on the coast. And finally, valley fog forms when moist air is trapped by the mountains into a valley." Summer finished reading the data.

The twins' phones received the uploaded information as the fog rolled in and completely covered the open field. The presence of a thick cloud didn't dampen anyone's spirits however, as plans for the next PB and J game were immediately underway.

"This is perfect," he thought. "This is such a thick fog that no one will see me when I nab those twins."

To make matters even better, he had overheard that, during the second game, the twins would be on a team with the robot squirrels so there wouldn't be any humans around to see him grab the twins. He was finally going to get them!

He could feel in his bones that he was on the cusp of the sweet revenge he had longed for. Cecil Sassafras would finally be paid back for what he had done to him those many years ago. He stood in the thick dark fog and smiled a thick dark smile.

"Ready or not, Blaine and Tracey, here I come."

## Chapter 13: The Forget-O-Nator

### *Competing Cycles*

This time around, they were the only humans on their team, and Blaine and Tracey were actually okay with that. It was the two of them, Ulysses S. Grant, and the robot squirrels. Between games, the Sassafrases had interacted enough with the robot squirrels that they were now comfortable with them, and they could see that the machines were truly back to normal. All of their kinks or malfunctions had been worked out by their inventor and caretaker, Ulysses S. Grant.

The twins continued to marvel at Summer Beach's lab assistant and resident inventor. Just like Uncle Cecil's sidekick, President Lincoln, Ulysses seemed to be capable of so much more than any normal animal should be. The twins gave Ulysses a high five, and then they rushed to get in place.

This second paintball game was a bit different than the first. This time, both teams were in the open field, but they were on opposite sides of the field from each other. A short flagpole had been placed in the very center of the field. It was each team's goal to get to the flagpole and hoist up their team's flag.

The two flags, however, were not normal flags, but rather small flags that had the translucent tablets affixed to them. The Sassafras team's flag and tablet had SCIDAT data on it about the nitrogen cycle. The other team's flag and tablet had SCIDAT information regarding the phosphorus cycle. The winning team of this game would be whichever team had their flag-tablet flying at the top of the flagpole by the end of the twenty minutes that were allotted for this game.

There was a marked starting line on each team's side of the field that no one could cross until Yotimo's whistle was blown, and that is where Blaine and Tracey were headed right now to wait. Their team's strategy for this game was to have Blaine and Tracey both sprint as quickly as possible toward the flagpole. The robots and Ulysses would fan out around the twins and shoot anybody from the other team they saw. The hope of the twins and their team was that they could get to the flagpole first, hoist their nitrogen cycle flag, and then defend the spot for the whole twenty minutes.

Both Sassafrases placed a foot on the line and listened intently for the whistle.

There it was! The starting whistle had been blown! Like shots out of paintball guns, the twins were off and sprinting.

The fog that had rolled in at the end of the first game was still present. It was like the whole field was covered with a thick gray blanket. It was a strange experience to run at full speed through fog like this, but the twins rushed forward into the haze. They estimated they could only see about fifteen to twenty feet in front of them. They were glad they knew this was a completely flat field because if it was unknown terrain, they would have needed to use a lot more caution than they were presently using. They just hoped they didn't accidentally run full speed into any members of the other team.

Blaine and Tracey raced forward with squirrels all around them, hoping this strategy would pay off. They really wanted to get to the flagpole first so that they could hoist and defend their team's flag for the duration of the game.

He had heard the whistle. It was time.

He emerged from his spot in the forest and stepped out into the open field. The fog was like a blanket. This was perfect. No one would be able to see him.

The sound of paintball guns firing their ammunition began to ring out. It sounded like chaos to him, and chaos was good. A chaotic setting would help him nab those twins.

"Oh, man!" Blaine shouted over to Tracey. "Somebody is already shooting! Let's hurry and get to that flagpole!"

Tracey nodded and ran forward with a little more tenacity. Almost immediately, she saw something, but it was not the flagpole. It was a…tree?

"Blaine!" Tracey hollered. "We ran in the wrong direction! Somehow we got to the forest, not the flagpole."

Blaine stopped and grunted because he saw that his sister was indeed correct.

"It's all this fog," he complained. "You can't see anything, much less keep your bearings straight."

"Let's not give up," Tracey responded, turning away from the tree they had reached. "Maybe the other team got confused in the fog as well. Maybe we can still get to the flagpole first. Let's go!"

The Sassafras girl raced back out into the field of fog with her brother right behind her.

"Wow, this fog is crazy!" Summer thought to herself. "Crazy fun, that is!" She loved that science was always available to explore. And this summer she was especially stoked that Blaine and Tracey were getting to explore so much of that science with her. It was easy to see that those cute little twins were now lovers of science. She was so proud of them. She continued to be amazed at how good they were at using their memories to retain all they saw and heard. All the scientific knowledge they were gaining would surely serve them well for the rest of their lives.

Summer giggled as she ran. She was having so much fun teaching the twins and getting to know them more and more—so much so, she almost felt like an aunt to them. The female scientist's heart excitedly skipped a beat, and her eyes glazed and began to look googly. How sweet would it be if someday she truly could be their aunt? That would mean that she and Cecil 'the Dream Machine' Sassafras would be married. She didn't know if it would ever happen, but her heart longed for nothing more. She loved that crazy redheaded scientist.

"Pop, pop, pop!" Paintballs were firing all around her. Summer smiled and got her head back into the game.

"Pop, pop, pop!" Paintballs were firing all around him, but he couldn't really see from which direction. Maybe all this fog wasn't going to serve his evil purpose well after all. He couldn't see a thing. How was he going to find those twins and grab them?

Pop! Splat! Pop! Splat! Pop! Splat! Ouch! Ouch! Ouch! He had just been hit three times on the top of the head by flying paintballs. Immediately, one of the robot squirrels scurried by him. This little machine was the culprit gunman. It chattered robotically as it passed him, almost like it was laughing. Then it disappeared into the haze.

Paint began to run down his eyebrowless face. He reached up in disgust and tried to wipe it away. He had endured so many pesky annoyances on his quest for vengeance.

"I can endure just a little more," he thought. "I just need to find those twins! I need to Forget-O-Nate them. Then all this science they have learned will be gone. And I will reign in vengeful victory!"

Oh, what was that? Was he now seeing the form of a person approaching him in the fog?

"This wasn't part of the plan," Tracey thought, as she sauntered slowly through the fog by herself. Somehow, she and Blaine had gotten separated. She had slowed down and was now creeping rather than running. She was watching for her brother, but she wasn't going to call out to him because she didn't want to give her position away to the other team.

What was that? She was approaching a form in the fog. Was that Blaine? The girl stepped cautiously closer. The form was not moving. Tracey suddenly stopped.

"This isn't Blaine," she mouthed silently. "This is . . . the flagpole!"

Tracey nervously smiled at the proposition of another possible paintball victory. "But where is Blaine?" she wondered.

The form he had spotted was still moving.

Slowly. Silently. Stealthily.

"This is definitely one of the twins," he thought to himself.

At this point, there was no place for him to hide, so he could either run off into the fog or he could move forward with his pernicious plan. He had run and failed too many times. He was going to move forward.

With a burst of wicked adrenaline, he lunged forward into the fog and tackled the approaching form in a monstrous collision.

Blaine felt like he couldn't breathe. He had just been hit. He reached up and felt his shoulder. Hey, wait. There was no paint there! The paintball must have glanced off him and not burst.

The boy sighed gratefully. A player was only out if the paintball actually burst open on them. Blaine was still in the game, but he was curious who had shot at him.

He crept forward, and all at once, he saw it—the flagpole!

Pop! Pop! Pop!

Paintballs started firing from all directions. The fog was still hanging low, but it was thin enough around the flagpole that Blaine could see what has going on. Tracey had just made a run for the flagpole, but at the last second, his sister had seen that the pole already had some tenants. It was being guarded by the Romigs!

Tracey was now backing up, but she was firing as she retreated.

At virtually the exact same time, two female voices both said, "I'm hit!"

Tina and Tracey both put their guns down, signaling they were out. Then they actually started laughing. It was good to know that while intense, this paintball jamboree was all for fun and enjoyment. Skeeter now started chuckling also, but he wasn't just laughing with the girls.

"I see you out there, Blaine," he said playfully. "Come on in and try to get me."

Blaine answered with a laugh of his own, and then he followed that up with a volley of shots from his paintball gun.

"You missed," Skeeter chuckled. "We've got our flag up, and by my estimation, there are only about five minutes left in the game. If you are going to make a move for the win, time is ticking away. Tick-tock. Tickety-tock."

Blaine didn't mind Skeeter Romig's playful banter at all. It actually reminded him a little bit of how his dad played around with him. The Sassafras boy switched tactics from direct approach to pullout. He was going to move back into the cover of fog and try to run around in a wide route to the other side of Skeeter. Maybe he could get the lumberjack-looking teacher from the other side.

Blaine just hoped he didn't run into Summer or any of the dogs on the way. Skeeter shot at Blaine a couple of times as the boy backed up, but Blaine made it into the cover of fog before his foe made contact. The twelve-year-old quickly ran in a wide half circle and then approached Skeeter and the flagpole from the other side.

Blaine tiptoed as quietly as he could back toward the all-important spot. He peeked through the haze, and it looked like Skeeter had his back turned to Blaine's new position. The Sassafras held up his gun and took aim. He pulled the trigger and fired one shot. Pop. Splat.

"Oh, man! Great shot, Blaine! You got me." Skeeter put down his paintball gun, and Blaine immediately moved in.

Sensing the urgency because of the ticking clock, the Sassafras quickly pulled down the phosphorus cycle flag and replaced it with his team's flag—the nitrogen cycle flag. He pulled on the pole's rope and hoisted the flag up to the top of the flagpole. Now the boy just had to defend this spot for however many more minutes or seconds were left in the game.

Paintball shots could still be heard out in the distance, but none were being fired here near the flagpole. The sound the Ramiqs were hoping for was the sound of Summer swooping in with a flurry of shots at the last second to win the game for the team. The sound the Sassafrases were hoping for was the sound of Yotimo blowing the whistle to end the game, giving their team the victory.

The Sassafrases ended up being the ones who heard the sound they were hoping to hear. Everyone cheered for Blaine, who had pulled out the win, and slowly but surely, the paintball players began to appear out of the thinning fog and started congregating around the flagpole. Skeeter ended up congratulating Blaine most sincerely and emphatically, even though it had been Blaine who had hit him with the shot that had knocked him out of the game.

It was also Skeeter who took the winning flag and tablet and read its SCIDAT data to the twins. "Nitrogen is an essential element for life on earth, and the nitrogen cycle explains how this element moves between plants, animals, bacteria, the air, and the soil," the schoolteacher read.

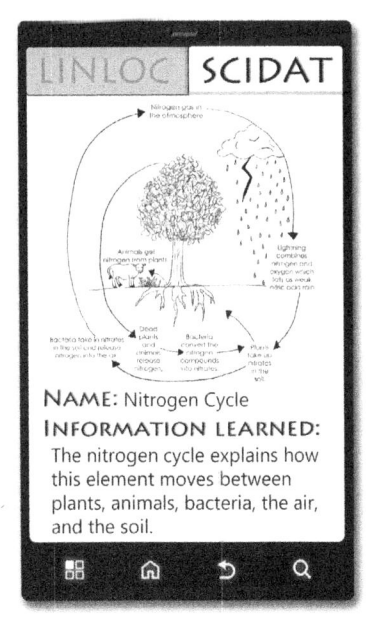

"Nitrogen gas is found in our atmosphere, and approximately seventy-eight percent of that atmosphere is nitrogen. Lightning combines nitrogen and oxygen into nitrogen-containing compounds, which then fall to the ground in rain. The nitrogen – containing compounds then soak into the soil. Bacteria convert those nitrogen compounds in the soil to make it usable for plants. Plants take up converted nitrogen from the soil, and animals get the nitrogen they need from eating these plants."

**NAME:** Nitrogen Cycle
**INFORMATION LEARNED:** The nitrogen cycle explains how this element moves between plants, animals, bacteria, the air, and the soil.

"When plants and animals die, they are broken down by bacteria known as decomposers. This process puts usable nitrogen-containing compounds back into the soil to begin the cycle again. The excess nitrogen-containing compounds are absorbed by special bacteria that convert the compounds and then release nitrogen gas into the air once more." Skeeter finished reading the information and then started looking for the 'upload' tab.

"Now, where is that button again?" he asked.

Tina reached over and showed him.

"Okay, now I see it," he said.

He pushed the tab and the SCIDAT immediately and wirelessly transported to the twins' smartphones.

Blaine and Tracey were happy. They had just received SCIDAT data. They had a couple of P.B. and J. victories under their belts. They were friends with the robot squirrels again. And they had new friends in the Romigs.

Yes, Blaine and Tracey Sassafras were happy, but they also felt a bit worried. Summer Beach had always been the one to upload SCIDAT data wirelessly to their phones here in Alaska (and once in Paris), but this time Skeeter and Tina had been the ones to help them. It wasn't that the twins didn't appreciate their new friends' assistance. It was just that they were concerned about why Summer wasn't there yet.

The twelve-year-olds looked around the field. The fog was starting to thin, allowing for a much wider range of sight. Yotimo and his sled dogs were almost at the flagpole, as well as Ulysses and all his robot squirrels. Everyone had arrived except the resident scientist.

Where in the world was Summer Beach?

### *Phosphorus Predicaments*

Even though he was the one who had dealt the blow, the impact of the collision had been so violent it had left him a little woozy. He had felt almost like he did in a landing location after light speed zip-lining. He had gathered his senses and then been met with a rush of confused yet elated emotions.

He was confused because the person he had tackled was not one of the Sassafras twins. He was elated because the person he had tackled was, in fact, Summer T. Beach.

She was lying on her back, sprawled out and unconscious. He had knocked her out when he had tackled her. Yet another location in which he had failed to nab the twins, but it looked like he was still going to be able to nab the favorite local expert instead.

Maybe abducting Summer was even better than abducting Blaine or Tracey. Maybe putting her in the Forget-O-Nator to wipe away her memory would cut at Cecil Sassafras's heart even more than if he did the same to one of those twins.

He snickered wickedly as he secured a harness and three-ringed carabiner to the unconscious scientist. "Time to zip-o-nate her to the forget-o-nator," he rasped.

"I don't know. Do you think maybe she went down to the lab?" Tina responded, after Blaine, Tracey, and Skeeter had all asked where Summer was. It had been a quarter of an hour since the paintball game had ended, and there was still no trace of Summer.

"I guess that's possible," agreed Skeeter. "It's definitely worth checking out, but why do you think she didn't even finish the paintball game?"

Everyone shrugged.

"So you guys know about the underground science lab?" Tracey asked the Romigs.

"Yep," Skeeter answered. "Summer showed us around the lab facilities a couple of days after we moved here. She told us that a lot of the research she does is top secret, but I guess she invited us down there so she could have a little human interaction. I assume it can get pretty lonely out here in rural Alaska."

"She also told us a little about all the fun you two are having this summer," Tina said. "Zipping around the globe and learning science face-to-face sounds magical!"

The twins nodded, confirming that it was indeed magical. They knew they weren't supposed to tell anyone about the invisible zip-lines, but they supposed Summer knew exactly what she was doing and was free to divulge information to whomever she saw fit.

Yotimo interjected into the conversation, not with words but by beckoning for everyone to follow him. Everyone did because no one questions a knife-wielding, stone-faced Eskimo. The big strong musher led the group of humans, animals, and robots back to the edge of the forest, where he stopped in front of a big tree. He pulled out his knife and jabbed it into a small slot hidden in the pattern of the bark.

The Eskimo twisted his knife and immediately a door opened up in the trunk of the tree. Yotimo stepped inside the doorway and disappeared down a ladder.

The Sassafrases and Romigs looked at each other with wide eyes. None of the four humans really knew what was happening, but they were going to follow Yotimo to find out. The Romigs let the Sassafrases go first and then they climbed down the ladder. They were followed by all the robots and dogs, and then Ulysses came down last. Upon reaching the bottom of the ladder, all saw they had reached a long hallway.

The twins smiled. They figured this hallway was a part of Summer's big underground science facility. Evidently, there were more entrances to the lab than just the spiral slide. The hodgepodge group followed the big Eskimo down the long lit corridor.

The straightaway changed into a series of sharp turns and then ended up at a closed metal door. Yotimo grasped the circular wheel on the door and cranked it hard to the left. The wheel creaked and spun and the metal door opened. Yotimo stepped through the door into another hallway that was longer, wider, and egg-shaped. The twins immediately knew where they were.

This was the hallway that led to Summer's main lab. They followed Yotimo at his brisk pace all the way to the end of the new hallway, where the musher hit a button on the wall. A big egg-shaped door began to open and bright light and anticipation spilled out onto the group in the hallway. Surely this is where Summer had disappeared to.

The large door opened all the way, revealing translucent floor to ceiling data screens, brightly-lit specimen tubes, and several tidy work stations. But there was no Summer Beach. The scientist was not there.

She slowly blinked her eyes open and immediately saw that she was no longer in the paintball playing field. What had happened to her? She halfway remembered running through the fog with her paintball gun, and then something . . . had jumped out of the haze and had . . . tackled her? Was that right?

She sat up from the spot she had been laying in and looked around. The room looked a lot like one of the rooms she had down in her underground science facility, with all kinds of monitors and computers, but this was not her room. She had never been here before. And what was that in the center of the room? Was that a port-o-potty?

Suddenly, a redhead flashed up on one of the monitors. Summer's heart skipped a beat. It was Cecil—Cecil the Studmuffin Sassafras. He was on the screen, happily bouncing around in his messy basement lab along with President Lincoln, his lab assistant. Oh, how she loved that wacky scientist, but why was he being broadcast on this monitor? Was this a live feed?

"Hello, Ssssummer," a deep snake-like voice sounded, interrupting Summer's questioning brain.

The frizzy haired scientist turned toward the voice and saw a big mean looking Man with No Eyebrows.

"If she's not up in the field and she's not down here in the lab, where could she be?" Tracey asked out loud, on the verge of fear-inducing worry.

"There's no telling," Skeeter answered in a shrug. "But look at the data screen. It has information showing about the phosphorus cycle. Isn't that the next topic you were supposed to study?"

The twins nodded.

"Maybe Summer left that information on the screen on purpose," Tina offered. "Maybe it was her plan all along to disappear and then have you hunt for your SCIDAT data."

"Maybe," both twins sounded very unsure.

The Sassafrases glanced over toward Yotimo, thinking that maybe he would have some kind of answer for them, but the Eskimo's face gave nothing up. They couldn't tell if he was worried or relaxed or mad or happy. They knew that Summer trusted this native Alaskan man. Surely, if something had gone awry and he knew about it, he would let them know.

The twins turned back toward the Romigs and the data screen.

"I guess let's go ahead and read and upload the data," Blaine said half-heartedly. He wasn't positive this was the right thing to do, but it was seemingly the only option they had.

Sensing the twelve-year-olds were a little down, Skeeter began reading the information on the screen in an upbeat manner. "Phosphorus is another essential element for life on earth," he read. "The phosphorus cycle explains how this element moves between plants, animals, water, and soil, but in all actuality, it is not a true cycle because a fair amount of phosphorus is lost along the way."

The twins felt a little lost without Summer giving them the SCIDAT on this leg of learning, but they did appreciate the effort and joy that the Romigs were displaying.

"Rock found in the ground contains phosphate, which is

a phosphorus-containing compound," Skeeter said, reading on. "When it rains, some of the phosphate leaks out of the rock into the surrounding soil and water. Plants absorb this phosphate and animals can get the phosphorus they need by eating these plants or by drinking phosphate-rich water. When the plants and animals die, decomposers break them down and release phosphates into the surrounding soil."

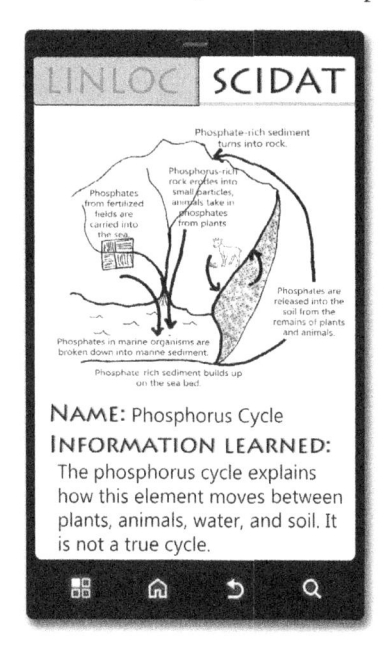

"In the ocean, much of the phosphates fall to the bottom, rendering them inaccessible to plants, which means they cannot be used to continue the cycle. This accounts for most of the 'lost' phosphorus in the phosphorus cycle."

Tina helped her husband find the 'upload' tab again, and the SCIDAT data was successfully transferred to the Sassafases' smartphones. The Romigs also helped Blaine and Tracey flip through the archive application and find perfect pictures for the water cycle, fog, the nitrogen cycle, and the phosphorus cycle. The twins then sent in all the SCIDAT data and pictures to Uncle Cecil's basement, where he could view it on the data screen.

"Well, I guess it's time to zip to the next location," Blaine said weakly. "Should we open up the LINLOC app and find out what, where, and with whom we will be studying next?"

"But Blaine," Tracey interjected. "How can we zip away without saying goodbye to Summer?

Summer carefully studied the man's eyebrowless face. She knew this man. She was sure of it. He looked a lot older and angrier than he used to, but she was positive that she knew him. He was an old classmate of hers.

The familiarity brought a smile to her face.

"Thaddeus," she exclaimed happily. "Long time no see! How have you been, buddy?"

The man took a step back, like he was shocked Summer knew who he was.

"You . . . you . . . remember me?" he asked. "You know who I am?"

"Well of course I do, silly," Summer answered. "After all, we were classmates, were we not?"

Some of the anger fell from the man's face, and his shoulders dropped a bit, but the softness only lasted a second before he brought his guard back up and became angry again.

"Well, then, you must also remember what Cecil Sassafras did to me!"

Summer thought back to her school days, but she couldn't come up with any kind of incident at the moment. Oh, wait! Maybe Thaddeus was talking about the time when . . .

"Yes, Cecil wronged me, and he will pay!" the Man with No Eyebrows thundered, cutting off Summer's reminiscing.

"Ever since that day so long ago in the classroom, I have been seeking revenge on Cecil Sassafras!" the angry man continued. "I have watched every move he has made. I have duplicated and perfected most of his best inventions, all for the purpose of vengeance!"

"His most prized invention is the invisible zip-lines. And the project closest to his heart is helping his niece and nephew learn science by using those invisible zip-lines to go and encounter science face to face around the globe. So if I could stop those twins, my thought was that I could crush all that was dear to Cecil, receiving my revenge in full. I successfully created my own three-ringed carabiner, which attaches to the invisible zip-lines. And I have chased Blaine and Tracey all over the planet. But those pesky twins have proven to be more resilient than I could have ever imagined."

This last statement about Blaine and Tracey brought a big proud smile to Summer's face.

"During their zoology studies, I left them marooned in the wilds of Africa. I enclosed them in a tomb. I spied on them with a robot hummingbird. I snuck into Cecil's basement and used his keyboard to jumble up their SCIDAT data, sending them to separate locations" the man yelled. "During their anatomy studies, I trapped them inside what I thought were indestructible boxes. I dressed up like a nurse and tried to thwart their learning. I hijacked Ulysses S. Grant's army of robot squirrels and used them to chase you and the twins all over your lab!"

Summer gasped and the man continued, "During their botany studies, I used a magical disappearing suit to spy on and chase them all over the place, but nothing I did worked! Those twins would not stop learning!"

At the mention of the twins' perseverance, Summer smiled again, even though the story that was coming together with all of this information was disturbing.

"Those twins kept learning, and I kept failing. I thought that is how it was always going to be," the man said, getting quieter. "That is, until today."

A wicked smile now formed on his face. "Out there, in the fog, I thought I had tackled one of the twins, but I hadn't. Instead, it was you. And when I saw you there, lying unconscious, I realized I had been going after the wrong target this entire summer. Instead of trying to stop those twins, I should have been coming after you!"

Summer gasped once again.

"The only thing that might be dearer to Cecil's heart than those twins is you." Summer's heart briefly stopped at being mentioned as dear to Cecil, as the man paused his monologue and walked over to the port-o-potty-looking contraption that stood in the middle of the room.

"This is what I call the Forget-O-Nator," the eyebrowless man said, pointing at the curious machine. "The way it works is simple. Put the subject inside and lock the door. Turn the one and only knob up to full capacity. After two minutes, the subject's memory will be completely wiped clean. My original plan was to nab each twin and put them inside this machine and wipe away their memories. This would, of course, include all the science they have learned this summer, but now that I have you . . ."

"Thaddeus! Please don't!" Summer said, pleading with her old classmate.

"I successfully grabbed Blaine earlier this week and put him here in the Forget-O-Nator, but somehow he managed to blow the machine up and escape. I spent several days repairing it, and I can assure you, no such mishap will occur today when I put you inside!" Thaddeus informed her.

Summer was not smiling. She was not clapping, jumping, or dancing. The scientist was actually crying as she was dragged across the floor against her will and shoved into the Forget-O-Nator.

"Thaddeus, please don't do this! Please stop!" Summer pleaded from inside the machine, but the door did not unlock and open. Instead a clicking noise could be heard as the one and only knob was cranked to full capacity.

THE SASSAFRAS SCIENCE ADVENTURES

Tracey pushed the button and the call went out.

"Howdy hooty," she heard the cheerful answer on the other end of the line.

"Hey, Uncle Cecil, this is Tracey."

"Hey there, niece-a-fras! How goes it? I just received your SCIDAT data and pictures about fog and the three different cycles! Well done!"

"Thanks, Uncle Cecil. Ummm, I just wanted to let you know that everything is going well, but we can't find Summer."

"You can't find Summer, you say?"

"Yes, that's right. We have looked everywhere we know to look, but she is nowhere to be found."

"Hmmm. Hum-diddly hmmm. That is curious."

"Well, what should we do, Uncle Cecil? Should we zip to the next location, or should we stick around for a little longer here in Alaska and find Summer?"

"Blaisey, you and Train should go ahead and zip zip zip away. I'm sure Summer is fine. She probably just slipped away and got caught up in a research project or something. You know how we scientists can be so absent-minded."

"Okay, Uncle Cecil. We will go ahead and move on. Your sure, right?"

"Sure I'm sure! Zip on and have fun! You guys are doing great!"

"OK, Uncle Cecil, goodbye."

Blaine had heard how the conversation and gone, so he quickly opened up the LINLOC app on his phone and took a look at the information it now held.

"Where to next, Sassafrases?" the Romigs asked with excitement.

"The Pacific Ocean!" Blaine answered. "Longitude 143° 38' 1", Latitude 26° 56' 20". We will be studying the topics of coral reefs, currents, oceans, and hurricanes, and our local expert's name is Billfrey Battaballabingo.

"Sounds exciting!" Skeeter resounded.

"It always is!" the twins said together.

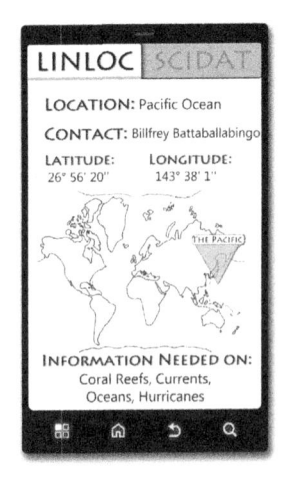

**LINLOC** SCIDAT

LOCATION: Pacific Ocean
CONTACT: Billfrey Battaballabingo
LATITUDE: 26° 56' 20"
LONGITUDE: 143° 38' 1"

INFORMATION NEEDED ON:
Coral Reefs, Currents, Oceans, Hurricanes

The fluorescent bulbs buzzed brighter and brighter, and as they did, the female scientist could almost feel the machine tugging at her brain. The Man with No Eyebrows laughed a wicked laugh on the outside of the machine.

As the woman cried tears of sorrow on the inside of the machine, she concentrated as intently as she could on the wonderful life she had lived. Her mind grasped for all it was worth at the events running through her mind. She knew this was the last time she would ever think these thoughts and relive these memories.

At one hundred and twenty seconds, the burning bulbs clinked off, and Summer T. Beach fell unconscious to the floor of the Forget-O-Nator. Her mind was now completely devoid of memory.

## Chapter 14: A Watery Landing... Again

### *Careening towards Coral*

It had happened again. The Sassafras twins had landed in water.

The two previous times they had zipped away from Summer Beach's lab in Alaska, they had landed in water at the next location. On their zoology leg, they had landed in the frigid waters of the South Atlantic. Then, they had landed in horse troughs in the United Arab Emirates while on their anatomy leg. And here again, while on their earth science leg, they had landed in a body of water after zipping away from Summer's lab.

This time, however, they had been halfway expecting to land in water, considering the fact that LINLOC had told them their location was the Pacific Ocean. They had landed with a splash. Thankfully, the water wasn't extremely cold and their heads didn't go under, so they weren't left gasping for breath. When the twins' vision and strength returned, though, they were shocked.

"Blaine, this is not good," Tracey yelped.

Blaine agreed, but he was at a loss for words.

"I figured we would land in water," Tracey continued. "But I thought the water that we would land in would have been by a boat, or an island, or a shoreline, or something. Blaine, there is nothing here! Nothing! What are we going to do? We can't tread water forever."

Blaine nodded but still wasn't speaking. He didn't have any ideas or answers. The Sassafras boy stared out over the calm, serene ocean. Somehow, this open water landing reminded him of the desert landing they had experienced a couple of days ago. The Gobi Desert had looked somewhat like it did here—nothing but blue above.

In all the blue of the Gobi, Blaine had almost felt like he was going to float up into the sky and disappear forever. But here, in all the blue of the Pacific, Blaine felt like he was going to sink down below the surface of the water and disappear somewhere down in the depths forever.

Blaine had to think of something to say to encourage her. That was a part of his duty as a brother, was it not? After all, he was the older sibling by five minutes and fourteen seconds.

Blaine opened his mouth, and what came out was, "Bunk bed."

Tracey looked at her brother like he had lost his mind. "Bunk bed?"

He pointed out across the water with wide eyes. Tracey followed her brother's pointing gaze. Her eyes opened even wider than Blaine's.

"Bunk bed!" Tracey also shouted like her brother.

THE SASSAFRAS SCIENCE ADVENTURES

Were they dreaming? Were they going crazy?

There was a wooden bunk bed cruising all by itself through the water. It was coming directly toward them.

The Sassafrases stopped shouting and started swimming, frantically flailing their arms and legs through the water in an attempt to elude the approaching bunk bed. It was no use; the double decker bed pummeled right into them.

The big wooden floating frame did not knock the twins under water, and it did not catapult them into the air. Instead, the drawer pull on the bottom of the bunk bed securely snagged them both by the loop at the top of their backpacks. The bed began dragging them speedily over through of the water. Blaine and Tracey looked at each other with mixed looks of panic and amusement as they skimmed along in the water.

Was this really happening? They had experienced some strange things this summer while learning science, but this current experience may have immediately jumped to the top of the list.

Realizing that this bed was the only ride for miles, Blaine and Tracey reached up to try to hoist themselves onto the bed. They had to wriggle out of their backpacks and do a back somersault, but they finally managed to get up onto the bottom bunk and out of the water. The bed continued to be pulled along at a surprising speed.

Tracey's mind processed three possible explanations of how a bunk bed could be speeding across the Pacific Ocean. There could have some kind of engine propelling it. It could be magic. Or, it could be being pulled by something underneath the surface—something strong and fast and big.

"Tracey!" Blaine shouted over the ocean spray to his sister. "Look! It's a fishing pole!"

Tracey looked where Blaine was pointing and saw what he was talking about. She realized that her third guess had been the right one. Just under the water's surface, a fishing pole could be seen. The pole was securely lodged crossways between and against the railing of the bunk bed. It was now almost acting like a grappling hook as it clung to the bed. The water was clear enough that Tracey could see the fishing line that was attached to the fishing pole, as well. It was being pulled completely taut with no slack and was angling down, disappearing into the water at a forty-five degree angle.

The line was way too long to see what was at the other end, and Tracey didn't know if she really wanted see what it was, anyway. Tracey shuddered. What kind of strong, fast, and big thing was pulling them?

As if his line of thinking was similar, Blaine shouted out a theory. "Tracey! Look at the line going down from the fishing pole. I bet somebody was out here deep sea fishing on a boat somewhere, and they hooked a big one. So big, in fact, that it pulled the rod and reel right out of their hands and into the ocean. Then the pole snagged this bunk bed, which for some reason was floating in the sea, and it started pulling it along, which in turn snagged us. That has got to be what happened, right?"

Tracey nodded, thinking Blaine's story was at least close. Besides, she hoped it was a big fish that was pulling them along and not some sort of weird and scary sea monster. Even if it was a fish, it would have to be huge and powerful to be able to pull this bunk bed this quickly. Tracey shuddered again, thinking about all of the strange creatures that lived under the surface of the sea.

"I mean, if humans haven't even been able to explore every place on land," Tracey thought. "Do I dare imagine what we would find if we were able to explore all the depths of the oceans of the world?"

Blaine was shuddering just like Tracey, as he gazed down as far as he could beneath the surface of the Pacific, thinking about what all might be down there. There were probably eels, snapping turtles, whales, sharks, octopus, and squid swimming around down there. There were probably sea snakes, sea horses, sea cows, and sea chickens deep down in the darkness below them, almost like a kind of water-filled farm.

"Wait," Blaine interrupted his own train of thought with another mental question. "Was there even such a thing as a sea chicken?"

There also had to be all kinds of fish. Fish with spikes and sharp teeth. Fish with weird eyes and fish that could light up. And probably even fish that could—what was that? Was Blaine imagining things or could he see something coming up toward him from under the water?

The Sassafras boy gripped tightly to the frame of the bunk bed. There was something coming up, and it was coming up fast.

"Tracey, look!" Blaine screeched, and pointed again with his terrified eyes.

Tracey saw it too and clung tightly to the bunk bed as well. There was nowhere for the Sassafrases to go. They were like bait on a hook for whatever huge creature was coming for them right now.

It came up fast, but then it stopped a few feet from the surface. It was the biggest creature either of the twins had ever seen. It seemed to be as big as the ocean floor itself. It was brightly colored and lumpy and looked like it was covered in…coral?

This was no attacking creature. This was a reef! Blaine and Tracey were now skimming over a coral reef. It had looked like something was coming up to get them, but it had just been this reef. Whatever it was that was pulling them on the other end of the line now slowed down a bit as they topped the reef, allowing the twins to peer down into the water.

The reef was teeming with beautiful color and life. The twins saw all kinds of cute little fish and other underwater creatures. They also saw some colors they were pretty sure they'd never seen before. This place was beautiful. So beautiful, in fact, they almost forgot they were stuck in a bunk bed being pulled across the ocean.

The two twelve-year-olds skimmed over the coral reef for a good long while and then were eventually pulled back out into deep open water. Whatever was pulling them was either so big or strong that it didn't even know it had a fishing line attached to it or it simply had unrelenting stamina.

Just then, the fishing line shot up toward the surface. Blaine and Tracey watched in awe as about forty feet in front of them, the biggest fish the twins had ever seen jumped out of the water, flew for a second, and then nose-dived back into the ocean. The other end of the fishing line that had snagged the bunk bed was hooked around the huge fish's spear-like nose. The Sassafrases' theory about being pulled by something under the water had been correct. It was the biggest, most amazing fish the twins had ever seen, but they still had no working theory as to why this bunk bed had been floating out in the Pacific.

The huge spearfish pulled the line back deep down into the water and tugged the heavy load onward. Slowly but surely, the fear the twins had felt about the mysteries of the deep faded away as they continued to be dragged along. Time passed and sonic lag began resting heavy on both of the travel-weary children. Their eyelids began to feel heavy. Both Blaine and Tracey began dozing off and lost track of time.

They were suddenly jolted out of their drowsiness, but it wasn't because they were sinking or because creatures were nibbling them. It was because their movement had abruptly stopped.

When their eyes opened, the twins immediately saw two things. They saw that the fishing line had gone slack, which hopefully meant the spearfish was now loose and swimming free. The second thing they saw was more curious. Their bunk bed joyride had come to an end next to a massive floating trash pile.

The pile spanned out about as far as the twins could see, and it was made up of all kinds of floating things. There were probably millions of plastic drinking bottles. There were rubber shoes, broken boat parts, basketballs, soccer balls, picture frames, playground equipment, car doors, various crates and boxes, long tangled tubing, broken toy pieces, and much more. The Sassafrases even saw a few other bed frames floating here in this crazy place.

"What in the world is this place?" both twins thought.

The twins looked beyond the trash pile and in all other directions. There still seemed to be no sign of any kind of land nearby. They were both itching to put some stable ground under their feet. Maybe they could find something in all this garbage that was firm enough to stand on.

Now that they weren't being pulled across the water at high speeds, they took the time to put their helmets, harnesses, and carabiners back in their backpacks.

Tracey noticed something. To the side of where they were, there was a series of rafts and planks that looked to be tied together, forming a sort of walkway through the trash. Tracey nudged Blaine and pointed to show him.

He nodded as they both looked in the direction of the makeshift walkway. As they looked, they saw that there weren't just flat rafts and planks, but also several small fort-like structures and a few curious looking statues. The statues were made out of odds and ends that had probably been found among the floating trash. They had been propped up and decorated by someone to look like standing human figures. There was a coatrack man, a ladder man, an actual mannequin, plus a broom man and mop lady that looked like they were holding hands. Things at this location just kept getting weirder and weirder.

The twins stepped out onto a log that led to the walkway, backpacks in hand. Blaine walked straight over to the mannequin, which was missing its right arm and its left leg, and held out his

## CHAPTER 14: A WATERY LANDING... AGAIN

hand for a handshake.

"Well hello there Mr. Mannequin," he joked. "My name is Blaine Sassafras and that over there, is my sister Tracey. You don't happen to be Billfrey Battaballabingo, do you?"

Tracey laughed at her brother's antics. That was, until the mannequin answered.

"Nope, that's not me," the mannequin responded. "But I do know the man.

Blaine stumbled back, stunned, as Tracey hunkered down in her spot, scared and afraid.

The mannequin then began to call out like a bird, "Kaw! Kaw!"

The Sassafras twins looked at each other in shock and confusion.

"Kaw! Kaw," the mannequin continued. Then it asked a question. "Where did the two of you come from? Kaw! Kaw!"

Guessing that the best option was to answer, Blaine stumbled through a response. "We . . . uh . . . zip . . . well . . . bunk-bed . . . spearfish . . . ummmm . . . most recently we came from cruising over a coral reef."

The mannequin did not reply and it did not "Kaw". Instead, a disheveled, bearded, shoeless, and sunburnt man dressed in threadbare shorts, a checkered tablecloth as a shirt, and a lampshade as a hat emerged from the small wooden fort directly behind the mannequin. He stood up and stretched, but then hunched over, took a wide stance, and shuffled over to where Blaine was in the strangest looking walk the twins had ever seen.

"Kaw! Kaw!" he shouted and flapped his hands and arms like birds wings.

Suddenly, he stood up straight, put one hand gracefully behind his back, gestured with the other hand in front of him, lifted his chin, and began pacing across the floating walkway like a lecturer.

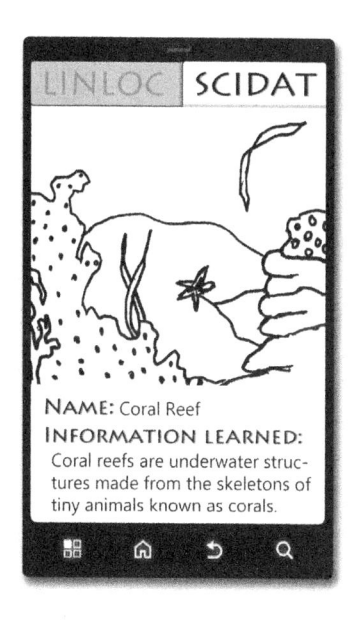

"Coral reefs are underwater structures made from the skeletons of tiny animals known as corals," he informed them, with a regal British accent. "The corals grow in colonies, and as older corals die off, new ones grow on top of them. Over time, the layers build upon each other to create a large structure. Many different varieties of coral growing and building together form reefs that are literally teeming with life. Coral reefs are home to more than fifteen percent of all fish species, even though they only cover one percent of the earth's surface."

"They form in the shallow waters of the world's oceans, which have plenty of wave action and temperatures ranging between sixty-two and eighty-two degrees Fahrenheit. The waves do not allow sediment to fall on the reef and they also bring nutrients and minerals to the animals that live there. Coral reefs are found throughout the warm tropical waters of the world, but the largest is the Great Barrier Reef just off the coast of Australia. It stretches over one thousand four hundred miles."

As soon as the man was done giving the scientific information, he hunched back over, shimmied into his strange walk, and started squawking like a bird again.

Blaine leaned over to his sister was and whispered, "This guy is bonkers!"

## CHAPTER 14: A WATERY LANDING... AGAIN

### *Drifting in Currents of Confusion*

Tracey agreed with her brother, their "expert" seemed to have a screw or two loose, but she wasn't sure they should zip away without their data. She voiced her thoughts to her twin. "Blaine, he is definitely the craziest person we've run into all summer, but he's also definitely our local expert. I don't think we have a choice here."

"You mean we have to stick around and deal with this guy?" Blaine asked.

Tracey nodded. "Yep. We've got to stay here and get the rest of our SCIDAT data."

Blaine knew his sister was right. He just really didn't want to hang out on a floating trash pile with a crazy birdman. The twins stood right next to each other and began slowly approaching the man, who was still flapping and squawking for all he was worth.

"It's pretty obvious he has dual personalities," Tracey whispered. "Maybe we can ask a question that will appeal to his refined scientific side."

"That's a good idea," Blaine replied, in a hushed voice.

"Mr. Battaballabingo," Tracey said. "What has created this floating mass of trash right here in the middle of the ocean?"

The man stopped his bird-like movements and sounds and stared at the twins. He was silent and still only for a moment. Then, he again straightened into a lecturer's posture and eloquently spoke.

"This is the Western Garbage Patch," he said. "It is a large area of spinning debris and trash. It was created by the currents that flow through the ocean and discarded non-biodegradable trash. The currents bring the debris to a stable location known as a gyre, where the trash is trapped. It spins slowly, and it doesn't move out of the area, which creates a vortex. Some of the debris remains on the surface, but much of it reaches closer to the ocean floor."

Billfrey Battaballabingo finished his statement and then reverted back into Birdman. He made his way over to the Ladder Statue. The twins assumed he was the one who had assembled all these strange human-like figures. Battaballabingo grabbed the paint tray that was attached by hinges to the wooden ladder and began flapping it up and down like it was the ladder-man's mouth.

"Hello, Blaine and Tracey Sassafras," Billfrey said, in yet another voice, as if he was the ladder itself. "It is so nice to make your acquaintance here in the Western Garbage Patch. My name is Cantankerous Carl, and I have been floating here for the longest time."

Billfrey lifted a plastic spring that was attached to the side of the ladder as if it was the ladder-man's arm. He then used it to point to other statues.

"That is Sticky Fingers Stevie." Cantankerous Carl referred to the coat rack figure who had a plastic rain jacket affixed to it. It also had a wide-brimmed rain hat attached to its top and one and a half rubber galoshes to its base.

He then pointed to the broom and mop that were 'holding hands.' Their hands were rubber gloves that were secured to the ends of lengths of PVC pipe, which in turn were attached to the broom and mop handles by dental floss.

"This is our lovely married couple, Mr and Mrs. Osodarling." The plastic springy arm then moved from pointing to the broom and mop and pointed to the mannequin.

"That over there is Ig, and then, last but not least, we have our resident marine biologist, Mr. Billfrey Battaballabingo," Battaballabingo said via Cantankerous Carl.

Billfrey was pointing the spring arm toward himself.

Blaine's mouth dropped open in utter disbelief at the lunacy he had just witnessed. Tracey, however, played along.

"It is so nice to meet you, Cantankerous Carl," she greeted kindly. "And you, too, Sticky Fingers Stevie, Mr. and Mrs. Osodarling, Ig, and Billfrey Battaballabingo."

Blaine slowly turned his head toward his sister. He couldn't believe she was continuing this converstation.

"Cantankerous Carl, is there anything else you think Billfrey Battaballabingo would want to share with us?" Tracey asked, continuing to speak to the ladder. "Like, say, maybe something more about the currents he briefly mentioned?"

"Of course he would," Carl replied. "He would be more than happy to."

Billfrey dropped the ladder's arm and mouth and began lecturing again.

"The water in the ocean is constantly moving due to currents," he said, in his own British accent. "The two main types of currents are surface currents and deep currents. Surface currents are due to the winds that affect the top of the ocean in a given area. These currents generally push water toward land and are responsible for the waves we see on the beach."

"Deep currents are due to the sinking action of cold water that comes from the north and south poles, which then drifts to the equator, warms up, and rises to the surface again. It then drifts back towards the poles, where it cools off, creating a cycle of rising and sinking water throughout the ocean. Deep water currents are also affected by the level of saltiness in the water, which is called the water's salinity. One last point that factors in is that the currents of the world's oceans are also produced by the spinning motion of the earth, which is a part of the Coriolis Effect."

Once again, as soon as the information came to an end, Billfrey Battaballabingo moved and squawked like a bird. As he did, the Sassafrases looked around again at the strange place where they found themselves. It was difficult to believe this much non-biodegradable trash had found its way all the way out here to the spinning Western Garbage Patch. It was also hard to believe this is where their local expert evidently lived. How did he survive out here by himself?

As the twins scanned the garbage patch, they saw a large number of birds. Maybe that was the reason Billfrey acted like a bird. Maybe he had been out here so long, with only these birds to interact with, that he was almost slowly becoming one himself. He must have created Cantankerous Carl, Sticky Fingers Stevie, Mr. and Mrs. Osodarling, and Ig so he could have objects other than birds to talk to.

Another question hit the twins at the same time: "What in the world did Battaballabingo eat?"

At that point, Billfrey stood up straight and began in his British accent, "And now, we come to the point in the day where we shall go to the fishing hole to get our daily victuals."

With that, he put his arms out and took off down a planked pathway across the garbage patch. Blaine and Tracey knew they had no choice but to follow. So with a sigh, they tightened up their backpack straps and begrudgingly pursued their Birdman expert.

It didn't take long to reach Billfrey's fishing hole, at least that's what the twins thought this place was. There was a hole in the garbage patch where there could see the water. Next to it was what appeared to be a recliner chair that was built out of old soda cans. Blaine and Tracey were looking around for a fishing pole, when all of the sudden their local expert pulled out what look like a large metal beak. He inserted a metal and rubber mouthpiece into his mouth before strapping the contraption to his head.

Then, to the surprise of the twins, Billfrey took a deep breath and stuck his head down the hole. Within a few seconds he was pulling up a decent sized fish with his metal beak. He sat back in the soda-can recliner and grabbed a glove with talons attached to it. The twins watched as the Birdman skillfully gutted the fish with his talon-glove.

"This guy really thinks he is a bird!" Blaine thought as his mouth stood wide-open.

Billfrey Battaballabingo stood up and screamed out, "Kaw, Kaw!" before taking off down the path from which they had just come.

Blaine and Tracey stood there too stunned to speak for a moment before Tracey broke the silence. "I guess lunch is going to be served back by his shack."

The twins started down the path once more, this time heading back towards the bunk-bed and bevy of garbage-statues.

When they got there, Billfrey had removed his beak and talon-glove. He had started a fire with some driftwood and was cooking the fish on what looked like the top of a discarded five gallon metal drum, and to the twins surprise, it actually smelled quite delicious. Billfrey's hair was slicked back, which made the twins hopeful that they would be conversing with the British side of their local expert, not the Birdman side.

"Would you like a spot of tea with your fish the afternoon?" the British-sounding Billfrey requested.

"Yes, that would be lovely," Tracey responded with her own attempt at a British accent, not skipping a beat.

Blaine, on the other hand, was just trying to keep up with all the changes. He managed to eke out, "Err . . . umm . . . yes."

Billfrey went into his shack and came back with three slightly-cracked tea cups and three Styrofoam plates. He dished up a bit of fish onto the plates and poured some brown liquid the twins hoped was tea into the cups. Their local expert handed a plate and a cup to each of the twins and gestured for them to sit down on a nearby plank. The twins took off their backpacks, set them down on the board, grabbed the food Billfrey offered, and sat down next to their respective packs on the plank.

He sat down on another plank, smiled at the twins, and took a sip of the tea-like liquid out of his cup, pinky up and all. Then, he set down his cup and without warning he cried out, "Kaw, Kaw" as he began pecking like a bird at the fish on his plate.

The twins watched the transformation with confused dismay—his back-and-forth performance was getting to be a bit trying. This local expert was nothing like the ones they had experienced in the past. They were beginning to wonder how and why their uncle had decided to send them to this location!

Blaine and Tracey shrugged simultaneously. If there was one thing they had learned on their summer zip-line adventure, it was to roll with the punches. Their stomachs growled loudly, so they dug into the fish on their plates using their hands. The twins weren't quite ready to copy their expert's beak-like eating habits yet.

As they were eating, the twins looked out onto their surroundings. Blaine stared straight ahead into the ocean, hoping that he could spot a boat or an island out there somewhere that would come rescue him off the crazy Birdman's garbage island. Tracey was staring out to their left, looking at all the creations Billfrey had made and admiring his creativity as she ate the delicious fish he had prepared.

Suddenly, a loud roar cut off the twins' thoughts. Their heads swiveled to their right to see a huge black claw burst out of a nearby wooden crate. The twins dropped their plates and squawked like birds themselves. They almost took flight in an attempt to jump away from the roaring box—maybe their local expert's bird-like antics were wearing off on them after all.

The twins landed safely on another plank and then turned back to look at the crate. The claw had recoiled back inside and the roaring faded to a growl.

Blaine and Tracey noticed that there were both Chinese characters and English letters written in red on the side of the crate. Of course, they couldn't read the Chinese, but the English said, "Taipei Zoo."

"Taipei Zoo?" Blaine questioned out loud. "How did a crate from the zoo get out here? And what kind of animal do you think is in there?"

The boy's last question was immediately answered as the crate exploded into a million splinters and out came a huge black bear. It roared again and barreled in the twins' direction.

The bear managed to make its way over the floating garbage pile, but just before it reached Blaine and Tracey, the animal misstepped, causing its front paws to punch through the patch and plunge into the water. The bear's chin plowed hard into the edge of the plank the twins were on, pushing it forcefully downward. The force catapulted the Sassafrases high into the air in different directions.

Tracey landed next to Battaballabingo, who was standing frozen like one of his garbage-statues on a raft. Tracey's backpack, however, had flown off in flight and landed elsewhere. Blaine landed with a smash and a crash onto another wooden crate that was labeled "Taipei Zoo."

He hit the box so hard he busted a big hole in the top of it. Uninjured, Blaine rolled off the crate and watched in whimsical horror as three blue-cheeked baboons climbed out of the crate. One baboon bounded off across the garbage. One grabbed Blaine's backpack and tossed it up into the air. One stared directly at the boy, opened its mouth, and let out the loudest and most awful screech Blaine had ever heard. Blaine stood where he was, staring face-to-face with the third baboon and screamed like a baby.

Meanwhile, Tracey Sassafras looked around and assessed the current situation unfolding here in this crazy place where a bunk bed had brought her and her brother.

Blaine stood, bawling.

Billfrey Battaballabingo stood bewildered.

A black bear and three blue-cheeked baboons had burst from boxes and were now bounding, bouncing, and barreling around, threatening to bite, beat, and bully the boy and girl.

Both Blaine's and Tracey's backpacks had been boosted.

"Boy, oh, boy," Tracey bemoaned. "What a burdensome bind."

## Chapter 15: The Threat of Thaddaeus
### *Oceanic Occurrences*

Thaddaeus was his name, and he was currently zipping and swirling through the Pacific Ocean. At first he had been a mere tropical storm, but they had upgraded him to a full-fledged hurricane—Typhoon Thaddaeus, to be exact.

"That's not good, President Lincoln. That's not good at all," the red-headed scientist mused out loud at what he had just heard the meteorologist say on TV.

Cecil Sassafras and his prairie dog lab assistant quickly scurried from the living room, where the TV was still buzzing with news about the powerful storm in the Pacific Ocean. They went into the basement, where the tracking screen was displaying two green dots representing his niece and nephew in the Pacific.

"Linc-dawg," Cecil said to his furry friend, as they looked at the two dots, "the invisible zip-lines have worked great-errificly this summer. They have successfully taken Blaine and Tracey all over the globe. But those two little science-learning machines have never had to use the zip-lines in hundred-mile-an-hour winds before. I am afraid that the invisible zip-lines might not have the capacity to work in the middle of a hurricane."

President Lincoln nodded his head and chattered as if he agreed.

Cecil continued. "By what I just saw on the weather, I don't think the winds of Typhoon Thaddaeus have quite reached the twins yet, but surely they will soon. I should probably call to warn them. Hopefully, they can get all their data gathered quickly and zip out of there before they are stuck."

President Lincoln chattered in agreement again. Cecil pulled his phone out of the pocket on his labcoat, flipped it up in the air, and caught it in his other hand. He found Blaine in his contact list and then pushed the tab to make the call. At the other end of the line the phone rang, rang, again, and then rang some more.

"Hmmm, that's strange," Cecil said. "What do you suppose is going on out there in the Pacific? Since Blaine isn't answering, let me try Tracey."

The scientist dialed up his niece, but her phone just rang and rang as well. "I wonder why they aren't answering," the scientist questioned toward President Lincoln.

The black bear and the three baboons were doing a fantastic job at breaking apart the network of rafts, planks, and forts Billfrey Battaballabingo had built and tied together as they chased the three humans around the garbage patch. Blaine had stopped his terrified scream and was now jumping to and from pieces of floating debris with agile elusiveness.

Tracey was being driven by pure adrenaline as she jumped and hopped and ran away from the seemingly rabid attackers.

Billfrey Battaballabingo was still acting like a bird, which actually seemed to be serving him well as he squawked and flapped away from the lost zoo animals. Cantankerous Carl, Sticky Fingers Stevie, Mr. and Mrs. Osodarling, and Ig all simply stood where they were.

As Tracey pushed forward, her mind wandered. "What was the story with these four animals, anyway? How had they ended up here? The crates they were in must have somehow fallen off the transport ship they were on and then gotten carried to the Western Garbage Patch by the currents like everything else here."

Tracey shook her head. "A rational explanation is not the priority here!" she reprimanded herself.

First and foremost, she needed to survive this harrowing game of tag. Then, if she and her brother could find and retrieve their backpacks, they could get hooked up to their harnesses, calibrate their three-ringed carabiners, and zip out of here. There was no longer a glitch in the zip-line system that prevented them from doing this without completed SCIDAT data. But she had yet to spot either backpack.

One of the blue-cheeked baboons had singled Blaine out and was relentlessly chasing the boy. The primate was better at bounding over the trash, but Blaine had managed to elude the baboon by being a quick decision maker. He wasn't second-guessing any of his steps. The Sassafras boy was going full steam ahead, maneuvering with agility over whatever obstacle he found in front of him. Right now, he was quickly stepping over a length of monkey bars that was floating on the surface. A misstep here would surely cause him to get caught.

At the end of the floating monkey bars was a large roll of bubble wrap. Blaine made it to the bubble wrap and bounded off of it with both feet. "Pa-pop, pa-pop!" the roll sounded as Blaine jumped from it.

"Uh oh," Blaine yelped, in his now airborne state. He was heading for an open ice chest, and he didn't think that would be a very good landing spot.

He tried to change his course in mid-air, but all he managed to do was land in the ice chest, hind-end first. Blaine was now wedged inside the chest with his feet, hands, and head sticking out.

'Pa-pop, pa-pop!' And now the baboon was jumping off the roll of bubble wrap, directly toward him.

Tracey didn't know why the bear had chosen to chase her—maybe it was the fish on her fingers. Or maybe she just exuded bear attractant. After all, she had been chased by a polar bear on their zoology leg . . . and now this. Tracey knew there was little time to sit around pouting or asking questions. She needed to keep running!

The twelve-year-old girl found herself in quite the predicament. If she continued to run over the floating debris, the bear would just keep chasing her. If she jumped into the water to swim away, she figured the black bear would just jump in and swim after her. Was there anything she could climb up onto?

Tracey jumped from an old drugstore sign to a white picket gate and then found herself back at the bunk bed that had brought her and her brother to this place. Tracey was excited. She had found something she could climb. The Sassafras girl grabbed ahold of the frame and climbed up onto the top bunk. She managed a small smile and looked back to see how close the pursuing bear was.

"Wait!" Tracey suddenly thought, as her heart dropped. "Bears are excellent climbers . . ."

By the looks of it, he had them thoroughly confused. His loud squawking and wild arm flapping had these two baboons' heads spinning. They weren't even chasing him anymore. They were sitting together on a floating door in front of him, looking at him as if he was crazy, which he was, kind of. What the two baboons didn't know is that Billfrey Battaballabingo had lured them to this very spot. He had spotted a wadded up net that he was going to use to trap these two.

As quickly as a hawk, the marine biologist reached down and grabbed the net. Then he tossed the tangled mesh at the blue-cheeked primates. It came down directly over the top of them, and before they knew what had happened, they were captured. They began screeching loudly and clawing at the net, but that only served to make them even more tangled.

One of those animals had come by and knocked one of his arms off, and he wasn't very happy about it. Cantankerous Carl, the ladder man, watched with his plastic saucer eyes as the four animals continued to chase the three humans around the Western Garbage patch. He didn't understand the silly shenanigans of these living breathing beings. All they were succeeding in doing was breaking apart the only home he had ever known. The raft he was on was in shambles and it had broken free from the rafts of his friends, Sticky Fingers Stevie and Ig. The two of them, at least, were still nearby. Mr. and Mrs. Osodarling's raft had been separated first, and the lovely couple was now floating off into the ocean all by themselves.

The darling couple must have been caught by a current because they were now so far out in the ocean they were just a dot on the horizon. He sure would miss them. Sticky Fingers Stevie and Ig were still fairly close, but they were starting to float away as well. Cantankerous Carl did not want to be separated from all his friends. He noticed that the wind over the Pacific seemed to be picking up a bit. He wondered if that meant there was a storm coming and if the wind was also affecting how he and his mannequin friends were drifting.

Another thing he noticed was that there was a backpack resting up against his bottom rung. Where had that come from? And, Sticky Fingers had a backpack floating with him on his raft too! One last thing the Ladder Man noticed was some sort of gray fin sticking out of the water, circling his raft.

Tracey clung, trembling, to the top of the bunk bed. The bear was now stepping from the drug store sign to the white picket gate. She had nowhere to go. The black bear would soon be upon her. Wait! What had happened? The bear had gotten his hind leg caught in the gate! And somehow, the gate had gotten tangled up in a heavy rope that was attached to a buoy. The bear growled and reached for the bunk bed, but it could not reach it. Nor could it reach Tracey!

Somehow, he had managed to pop his body free from the ice chest, like a cork out of a bottle, just before the baboon had landed on him. And to make the situation even better, the screeching animal had then landed in the ice chest. Blaine had been able to close and lock the lid on top of it, effectively capturing his pursuer.

Blaine sighed a big sigh of relief. Then he stood up straight to see what the situation was with his local expert and his sister. About thirty or forty yards away, he saw the two standing together, with no animals chasing them. It looked like the black bear and the other baboons had been subdued as well.

Billfrey Battaballabingo was doing his usual thing and was acting like a bird, but Tracey looked more than a little worried. She was waving at Blaine and imploring him to come over to where she and Battaballabingo were.

The Sassafras boy made his way to his two companions' spot as swiftly as possible. When he got closer, he could hear his sister's anxious voice.

"Blaine! Hurry up! I spotted our backpacks, but they are on rafts that are floating away!"

"You spotted our backpacks?" Blaine responded.

"Yes! But we are going to lose them if you don't hurry. Everything is breaking apart and floating away fast!"

Blaine jumped from a grandfather clock to a chest of drawers and then to a church pew. He sprinted down the length of the pew and then dove out over open water toward the large raft that both Tracey and Billfrey were on.

Blaine's legs slapped hard against the water, and his torso hit the planks of the raft with a thud, but he had made it. Tracey smiled in relief and then helped her brother climb all the way up onto the raft. She patted him on the back, and then she pointed out into the ocean before them.

"Look! See the raft Cantankerous Carl is on?"

Blaine nodded.

"One backpack is on his raft," the girl confirmed. "And the second backpack is on Sticky Finger Stevie's raft."

Even though both the rafts were fairly far away from them at this point, Blaine could clearly see the two backpacks.

"Well, what do you suppose the fastest way to get over there is?" Blaine asked. "Should we swim over? Or should we try to paddle this raft in that direction?"

Tracey wasn't sure what the best option was. She turned back toward Billfrey the Birdman to see if he might offer any suggestions, and she was surprised to see that the marine biologist was not impersonating a bird. Rather, he had a hand elegantly placed behind his back. He was gesturing with the other hand in front of him. His chin was up, and it looked like he was about to begin a lecture.

"Oceans cover nearly two-thirds of the earth's surface," he said in a refined voice. "And all the world's oceans are joined, enabling the currents to carry water from ocean to ocean."

He was, indeed, lecturing again. It was information the twins wanted and needed to know, but they also needed their backpacks. So while Battaballabingo lectured, the twins each stuck an arm in the water (Blaine on the raft's left and Tracey on the raft's right), and they started paddling out toward the escaping raft of Cantankerous Carl.

"The floor of the ocean is similar to the earth's surface," Billfrey continued. "It has valleys, mountains, hills, and trenches. In general, the seabed slopes downhill gradually away from land to form a large shelf known as the 'Continental Shelf.' It then drops quickly away to the deeper part of the ocean at the 'Continental

> **LINLOC SCIDAT**
>
> **NAME:** Oceans
> **INFORMATION LEARNED:** Oceans cover nearly two-thirds of the earth's surface.

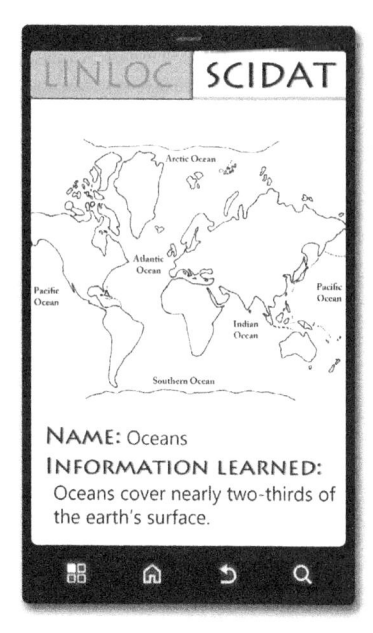

Slope.' The deepest parts of the ocean are called the 'Abyssal Plain.' The very deepest point of the ocean is the Marianas Trench, which is located right here in the Pacific Ocean, off the coast of the Philippines."

The twins felt like they weren't going fast enough to catch the rafts, so they both stuck the entire lower halves of their bodies in the water off the back of the raft and started kicking for propulsion, which seemed to work a little better than paddling with their arms.

"There are five main oceans," their strange yet brilliant local expert continued. "The Pacific is the largest, and that is the ocean we are in right now. It covers nearly a third of the earth's surface. It is located between North and South America, Asia, and Australia."

"The Atlantic is the second largest. It is located between North and South America, Europe, and Africa. The Arctic Ocean is located near the North Pole. The Southern Ocean is located near the South Pole. And then finally the Indian Ocean is located between Africa, Asia, and Australia. There are also many seas around the globe, but a sea is much smaller than an ocean and is typically partially enclosed by land. Seas are usually found where oceans and land meet."

Billfrey Battaballabingo finished his spiel about oceans, and then back to being a bird it was. The twins were pretty sure their local expert had experienced a mental crack at some point in his life. No normal person they had ever met acted like he did. However, even though he was weird and awkward, both twins were learning to like and trust the man. Uncle Cecil and the invisible zip-lines had connected Blaine and Tracey to all different kinds of scientific experts around the world. And they had all proved their worth to the Sassafras children.

### *Typhoon Thaddaeus*

As they got farther into the open ocean, Blaine noticed it was getting windier and that now there were waves whereas before there had been none. Tracey also noticed that the sun had disappeared behind some clouds and the sky was growing darker as if night was approaching. But the thing the twins were focused on was getting closer and closer to the runaway rafts they were chasing.

Blaine and Tracey slid their bodies all the way into the water and kicked as powerfully as they could with their feet. They each paddled with one hand and pushed the raft with the other. They figured if they could just make one last energized push, they could reach the closest free-floating raft. The Sassafras twins stroked their arms hard through the water and churned their legs for all they were worth. They had to get their backpacks back from Cantankerous Carl.

They focused in on the ladder man, who was now about fifty feet away, rising and falling on the growing waves.

Forty feet, they kept kicking.

Thirty feet, they kept stroking.

Twenty feet, they kept going.

Ten feet, they were going to make it!

Then, they noticed something. It was a gray fin that had briefly skimmed the surface of the water. Prickly pangs of excruciating fear shot through the twelve-year-olds' hearts and minds. Had they just seen a shark?

Swish! Swipe! Something underwater had grazed across one

of Blaine's legs.

Bump! Slam! Something strong had hit one of Tracey's arms.

As quick as whips, the twins each shot up out of the water and landed on the large piece of wood they had been pushing. They huddled together as closely as they could at the very center of the raft.

Billfrey was standing there, without even the smallest hint of angst on his face. The Sassafrases, however, stared at the water around them with looks of horror on theirs.

"Ahhh!" Tracey screamed. "I saw it again! The shark fin came up right there! On this side of the raft!"

Blaine had seen it, too, but now he was seeing something even more terrifying on the other side. There was a second shark fin swimming by! Now he saw a third . . . and a fourth!

"We're surrounded!" the Sassafras boy shouted.

They had gotten so close to reaching the first of their two wandering backpacks, but now they were encircled by sharks.

Cantankerous Carl bobbed silently just a few feet away, with one of the backpacks resting against his bottom rung. But there was no safe way to get to him. They obviously couldn't swim over. It was too far to jump. They couldn't zip out of the situation because all the necessary gear to do that was, of course, in their backpacks. They couldn't stick any of their limbs in the water to paddle over. What were they going to do?

To make matters worse, the raft they were on now began to drift away from Cantankerous Carl. The twins' hearts sank. It looked like their destiny for today was to be castaways lost at sea on a raft with a crazy man and surrounded by sharks.

All at once, Billfrey Battaballabingo reached up and took off his lampshade hat. Resting on top of his head, was a piece of coiled rope. He grabbed it and let it unfurl, revealing that the length of rope was actually a sort of lasso. The marine biologist tossed the rope into the air. It sailed over the shark-infested water toward the other raft, and landed around Cantankerous Carl's "head."

Battaballabingo carefully pulled on the rope, bringing the two rafts together. The twins were more than pleasantly surprised at what had just taken place, but Billfrey was not yet done. He unhooked the lasso from Cantankerous Carl and took aim at Ig, whose raft was the next closest. He successfully lassoed the one-armed, one-legged mannequin and managed to pull his raft in to join them.

Now for Sticky Fingers Stevie—the coat rack man's raft was uncomfortabley far away. Blaine and Tracey weren't even sure if Billfrey's lasso was long enough to reach it. Nonetheless, the Birdman was going to give it a whirl.

The sharks continued to circle. The marine biologist eyed his target.

The sharks continued to circle. Billfrey twirled the rope around in his hand.

The sharks continued to circle. Battaballabingo let the length of rope fly.

The sharks continued to circle. The lasso reached for the coat rack. The sharks continued to circle. The lasso…hooked over the top of Sticky Fingers Stevie!

Billfrey Battaballabingo had done it! The marine biologist now gently pulled the fourth raft in to join the other three. He then securely tied all the rafts together using the length of rope. The sharks continued to circle, but they didn't seem like as much of a threat anymore. As the twins bobbed up and down on the waves with their four strange friends, their anxiety began to diminish.

They had retrieved their backpacks. They were on a sizeable raft. They could zip away if they wanted to, even though they had yet to get their SCIDAT data about hurricanes. And their local expert seemed skilled enough to handle any situation this ocean might throw at them.

Billfrey Battaballabingo put his lampshade hat back on and then stepped over next to his handmade ladder man. He grabbed the paint tray mouth and again began speaking as Cantankerous Carl.

"Well, Blaine and Tracey Sassafras, that was a close call! I almost got separated from all my friends, but thankfully Billfrey Battaballabingo was able to keep us all together. Everyone except for Mr. and Mrs. Osodarling, that is. They floated off into the wide blue yonder far beyond eyesight. I sure will miss them. They were such a lovely couple."

This time, Blaine wasn't looking at his local expert like he was crazy. Instead, he joined in conversation with him.

"Cantankerous Carl, what do we do now? How do we get rescued? How do we find land? What will we eat and drink if we get stuck out here? You can't drink seawater, can you?" Blaine asked.

Blaine knew he and Tracey could zip away if they wanted to, so he wasn't really worried for himself or his sister. He really was asking these questions about the plight of Billfrey, whom he was worried about.

"So many questions, Blaine, so many questions," the ladder man responded, with the paint tray flapping up and down. "I will answer your last question first and then your first question last. Then your second, third, and fourth questions not at all."

Blaine now asked his sixth question. "Huh?"

"No. You can't drink seawater," Cantankerous Carl went on. "At least you shouldn't. As a matter of fact, most of the world's water you shouldn't drink."

Tracey now added a, "Huh?" of her own.

"The earth has three main types of water," Carl continued. "Seawater, brackish water, and freshwater. Seawater has a high concentration of salt and is typically found in the world's oceans. Brackish water has less salt than saltwater but more than freshwater, and it is typically found where freshwater sources meet the ocean such as in deltas or estuaries. The Baltic Sea is the world's largest brackish water source. And then there is freshwater. It has little to no salt and is typically found in rivers, lakes, ponds, streams, and wetlands."

A drink of freshwater sounded tantalizing to the twins right now, and the temptation to zip away unfinished with this leg was hanging in their minds. But Sassafrases never give up, no matter what. Right now, there would be no drinking and no zipping. Blaine and Tracey would stay in this location until they were finished with their task, but before that happened it looked like they were going to get some shuteye. Because the local expert that held the information they needed in his mind was suddenly out cold.

Billfrey Battaballabingo had just been using Cantankerous Carl to talk about different kinds of water. And then, after he'd put the period on his last sentence, the marine biologist had laid down on his back on the raft and immediately fallen asleep. The man was already snoring—what an extravagantly strange local expert.

Blaine looked at Tracey and shrugged. "Well, I guess we might as well get some sleep, too."

Tracey wasn't too thrilled about sleeping on a bobbing raft in the middle of the ocean surrounded by sharks, but Blaine was right. Sleeping was the best, and really only, thing they could do right now. The twins lay down on their backs as close to the center of the big raft as possible, and before they knew it, they were both snoozing underneath the nighttime sky as the waves of the ocean rocked them.

Tracey dreamed she was competing against the Avargaloin giant in a water bucket-filling contest. She was actually beating the giant until he turned and started grabbing her buckets and pouring them on her. "Hey, that's cheating," Tracey yelled in her dream, but the giant kept drenching her.

Blaine dreamed he was a grandfather clock and that he was playing paintball. All of a sudden, a blue-cheeked baboon jumped out of the forest and dumped an ice chest full of water on him. "Hey, that's cheating," Blaine yelled in his dream, but the primate grabbed another ice chest and heaved the water from it at Blaine.

"Hey!" both twins yelled and sat up from sleeping. "What's with all the water throwing?"

"It's a typhoon, or hurricane as they are called in the western hemisphere," they both heard a British voice say.

The Sassafrases both opened their eyes from sleep. They were completely soaked and water was still coming at them from every direction. They saw a ladder man, a coat rack man, a mannequin, and a real man who was wearing a lampshade for a hat. The real man had his chin tilted up and was giving a lecture. As he talked, the twins remembered where they were.

"Hurricanes are huge storms that can be hundreds of miles wide, with lots of rain and damaging winds," their local expert was saying. "They usually only form

**NAME:** Hurricane
**INFORMATION LEARNED:** Hurricanes are huge storms that can be hundreds of miles wide, with lots of rain and damaging winds.

during the hurricane season, which is June first to November thirtieth, in warm and wet conditions like those found in the tropics. The ocean water needs to be eighty degrees Fahrenheit or warmer for a hurricane to form."

The twins' gazes now expanded out from their raft to the ocean around them, and their hearts were immediately filled with fear. The waves around them were no longer just rolling waves but were now monstrous waves that were rising and falling all around them at heights much larger than their makeshift raft. The wind was whipping and howling and rain was beating down in torrents. They were in the ocean during a hurricane, and they were scared.

Billfrey Battaballabingo, however, was as cool as a sea cucumber. "The strong winds of a hurricane spiral around, creating an eye at the center of the storm where things are calm," he said, like he was in a college classroom addressing a crowd of students, not on a tied-together raft in the ocean and addressing two shivering twelve-year-olds.

"When the winds hit land, they can cause major damage and flooding," Billfrey continued. "However, when a hurricane does hit land, its winds begin to weaken. In the Atlantic Ocean, the Gulf f Mexico, and the Eastern Pacific Ocean, these storms are called hurricanes. In the Western Pacific, they are known as typhoons. And in the Indian Ocean, the Bay of Bengal, and near Australia they are called cyclones."

Battaballabingo paused and switched the hand behind his back to his front, and the hand in front to behind his back. "We are currently only on the very edge of this storm, so we can expect to see bigger waves, stronger winds, and heavier rain."

That is not what Blaine and Tracey wanted to hear. All at once, seemingly out of nowhere, a huge ship appeared in the ocean in front of them. It was still a decent distance away, but it was bursting, bow first, through the rain, rising and falling on the huge waves as it barreled in their direction. It was an old looking ship, made of wood. It had three masts, crows' nests, and huge sails.

The twins would have thought it might be a pirate ship except for the fact that it was brightly painted in blue, orange, and white. But it did have cannons protruding from holes on its upper sides—maybe it was a pirate ship. But there were life rings hooked around the end of each cannon—maybe it was a rescue ship.

The twins were confused. Should they be hopeful or scared? The ship's flag now came into view, and it only added to the confusion. It was a black flag with a skull and crossbones on it, but it also had a red cross on it—a red cross with the paramedic symbol of snakes wrapped around a pole. Were they about to get blasted out of the water or saved?

The big ship came up and rested right beside them. The twins stared up at it, speechless. Battaballabingo, however, squawked like a bird.

One by one, burly, bearded, husky, rough-looking men began to peer over the ship's edge down toward the raft. They all looked like pirates, except for the fact that their clothes were blue, orange, and white, just like the ship. In the middle of the bearded heads, a tiny pale head with a wispy almost invisible beard poked out.

Wait! Blaine and Tracey had seen this skinny little guy somewhere before. They knew him! This wa—

"Well, hidee-ho, there, dear buddy friends," the little man said, in a squeaky little voice. "It seems as though you are lost at sea, bobbing hopelessly at the mercy of the waves on your little raft, but have no fear! We are here to rescue you."

Blaine and Tracey heaved sighs of relief. That was what they wanted to hear.

"My name is Peach Beard, and I am the leader of this ragtag group of Pira-medics."

"Pira-medics?" the Sassafrases questioned out loud.

"Yessirree, dear buddy friends," Peach Beard confirmed. "We are the Piramedics."

Both twins remembered Peach Beard and his friends from their run in with him off the coast of South Georgia Island while studying zoology.

"But I thought you were the P.R.O. Pirates?" Blaine questioned.

"Yeah, the Piracy Resurgence Organization," Tracey added.

Peach Beard chuckled. "Well, I suppose we are pirates at heart and always will be, but now what fills our sails with wind is sailing the high seas looking for poor souls to rescue."

"Yo ho!" all of Peach Beard's men shouted.

"How did you go from being P.R.O. Pirates to Pira-medics so quickly?" Blaine asked.

"Well, we got official bona fide genuine certificates on the internet," the wispy bearded man said with a smile. "We are now certified to rescue up to six people at a time from perilous oceanic situations and can perform first aid, CPR, and psychoanalysis."

A big wave swished and nearly capsized the raft.

"Looks like it's time to rescue you," Peach Beard announced from the big blue, orange, and white Pira-medic boat.

Blaine and Tracey nodded in drenched agreement. The bearded Pira-medics jumped into action and began lowering down one of their men with a stretcher. When the pira-medic got to the raft, he quickly grabbed Ig and put him on the stretcher and then signaled for his cohorts to hoist him up. Blaine and Tracey looked at each other, dumbfounded. What were these piramedics doing? Why were they rescuing a mannequin?

"Oh, no," Peach Beard said from the deck. "It looks like he is badly injured and not breathing! Let's get him up here quick!"

Ig was hoisted all the way up, taken off the stretcher, and then attended to by a host of pira medics. Thankfully, Peach Beard's men immediately began sending the stretcher back down. Though they hadn't been first, at least now the Sassafrases knew they were going to get rescued.

Or were they? The twins watched, utterly flabbergasted as the silly scallywag pira-medics next rescued Cantankerous Carl and Sticky Fingers Stevie. Why were they rescuing the fake garbage-patch-people before the real living, breathing people? After all the mannequins were rescued, the twins watched as Billfrey Battaballabingo was hoisted up in the stretcher.

Blaine turned to his sister. "Even if they are counting the three mannequins as real people, at least there will still be room for us."

"You mean because they are certified to rescue six people?" Tracey asked.

Blaine nodded.

But any optimism they gained was immediately crushed as Peach Beard called out, "Last man, coming up!"

"What do you mean, last man?" Blaine questioned. "You still have room for two more people! And besides, three of the 'people' you just pulled up aren't even rea—"

"Sorry, two little buddy friends," Peach Beard apologized with his ever-present smile. "I wish we could rescue the two of you as well, but rules are rules, and we already have our six."

"But you only brought up four," Tracey shouted up. "You still have room for two mo—"

The Sassafras girl stopped her statement as she now saw two new individuals leaning over the railing being held by Billfrey Battaballabingo, who was now safely on board the big ship. It was the broom man and the mop lady, Mr. and Mrs. Osodarling! They must've been rescued from the open ocean previously by the pira-medics.

The Sassafras twins were at an utter loss for words and felt immobilized by their bewilderment. They watched, completely aghast, as Peach Beard, the Piramedics, Billfrey Battaballabingo, and the five mannequins sailed off without them.

## CHAPTER 16: QUICK! TO SWITZERLAND!
### *Grim Groundwater*

The waves rose and fell higher and taller. The wind and rain beat harder and faster. The Sassafras twins clung to their raft.

They had already texted in the needed SCIDAT data to Uncle Cecil's basement, and now they were sending in the four pictures. After they were done, they would open up LINLOC, see where the next destination was, and zip out of this howling and harrowing place.

The 'SEND' button was pushed. The pictures went out. The LINLOC icon was pushed. The application opened up. The next location was Bern, Switzerland. Their local expert's name was Evan DeBlose, and their topics of study were groundwater, waterfalls, rivers, and lakes.

"Okay!" Blaine shouted to his sister over the thundering of the storm. "Let's calibrate our carabiners and zip out of here!"

Tracey was all in for that. Each twin turned the longitude ring on their carabiner to 7° 27' 29", the latitude ring to 46° 56' 50", and then they let them snap shut. Each carabiner found the desired invisible zip-line and pulled the twins up into the air a few feet off of their raft. They would hang here for approximately seven seconds, and then they would zip off at the speed of light on their way to Bern, Switzerland.

Blaine and Tracey looked at each other with relief, glad to be exiting the open water of the Pacific. But their relief was short lived, because seven seconds ticked by, and then fourteen, and then twenty-one, and still they had not taken off. They were, however, bouncing around as they hung there, and the unseen lines they were attached to were being whipped around and pulled or pushed at.

Tracey was about to ask her brother what he thought was wrong, when all of the sudden their carabiners unclipped from the lines and they fell back down to the raft.

"Blaine! What happened?? Do the invisible zip-lines not work anymore?"

Blaine just shook his head like he didn't know. "We put the correct coordinates in right?"

"I'm pretty sure we did," Tracey responded.

"Let's try again," the Sassafras boy said.

THE SASSAFRAS SCIENCE ADVENTURES

The twins again calibrated their three-ringed carabiners, and were very careful to make sure they were set to the right coordinates. They hung in the air for nearly half a minute and were met with the same outcome.

"Maybe it's the typhoon," Blaine offered, in alarm. "Maybe the invisible zip-lines aren't working because of the powerful winds!"

The Piramedics had left them.

Their local expert had left them.

The zip-lines weren't working.

Their raft was breaking apart.

The storm was growing more ferocious.

What were Blaine and Tracey going to do? They were hopelessly marooned.

Suddenly, seemingly out of nowhere, a rope ladder appeared in front of the twelve-year-olds. It had unfurled from the sky and it now dangled right there in front of them. Blaine and Tracey looked up and saw that a huge bird had flown in to rescue them; a bird with rotating blades that was shouting their names.

Wait! This was no bird! This was . . . the heliquickter—the super-fast helicopter that Ulysses S. Grant had invented for Summer Beach! Their favorite local expert was here to save them! They could hear her shouting their names right now.

"Blaine! Tracey! Hurry! Grab the ladder!"

The boy and girl were more than happy to obey the imploring command. Tracey grabbed a hold of the rope ladder first and began climbing up, with Blaine right behind her. The swinging and swaying ladder would have been hard enough to climb in sunshiny weather, but the stormy weather made it that much more difficult.

They had no idea how Summer had known that they needed rescuing or how she had pinpointed their location, but none of that much mattered. They were rescued and they were happy.

"Okay, you guys. Hold on tight and keep climbing up!" the voice shouted down. "We have to fly out of here! The storm is way too strong for the heliquickter!"

The flying machine broke from hovering to full force flying. The rope ladder changed from a vertical angle to more of a horizontal one, and the rain of the storm changed from feeling like drops hitting them to feeling more like small rocks hitting them. Nonetheless, both twins eventually made it to the top.

Tracey reached up. The woman was waiting their to help pull the twins up into the heliquickter. But as Tracey reached out, she saw that . . . it was not Summer.

"Oh Tracey! I am so glad you are safe!" Tina Romig pulled the Sassafras girl safely into the aircraft.

"Tina?" Tracey questioned. "But I thought . . ."

Blaine clambered up into the heliquickter behind his sister and immediately had the same confused look on his face. Skeeter Romig, who was there too, quickly pulled up the rope ladder and closed the heliquickter's side door.

"Your Uncle Cecil called us and told us about the typhoon," Tina told them. "He was afraid the zip-lines wouldn't work if they got crossed up in such a powerful storm."

"He was right about that," the twins said together.

"He tried to call each of you," Tina continued, "but he said that neither of you answered."

The twins pulled out their smartphones, scrolled over, and saw that they did indeed have several missed calls from their uncle.

"We were watching you on the tracking screen down in Summer's lab," Skeeter said. "so we knew right where you were. We decided, along with Cecil, that we had to come get you using the heliquickter. But even this heliquickter, as powerful as it is, doesn't have the ability to fly inside a typhoon, so I'm glad that you two were just on the edge of the storm."

Blaine and Tracey couldn't believe that they were still only on the edge. It sure had felt like they were right in the thick of it.

"Since the heliquickter is more than twice as fast as a normal helicopter," Skeeter continued. "We were able to speed down here to where you were in the Pacific, from where we were in Alaska at the northern end of the Pacific. Thanks to Yotimo's skilled piloting, we made it in time to sweep you up before Thaddeaus did."

"Thaddeaus?" Tracey asked.

"That's the name of the typhoon," Skeeter explained.

"Yotimo?" Blaine asked.

"Yep, he's flying the heliquickter," Tina said, pointing to the front of the aircraft.

Sure enough, there was the big strong silent Eskimo up there serving as pilot. Was there anything this man couldn't do?

"But what about Summer?" Tracey asked concerned. "Have you guys found her yet?"

The Romigs looked concerned as well, and shook their heads "No". The Sassafrases were very grateful for their new friends, but they were also very worried about Summer T. Beach.

When Yotimo had flown the heliquickter safely out of the danger zone, the twins sincerely thanked their three heroic friends. The twelve-year-olds said their goodbyes, re-calibrated their carabiners to the Bern, Switzerland coordinates, and let them snap shut. This time, everything worked like it was supposed to and the twelve-year-olds zipped away from the heliquickter, through swirls of amazing light, on their way to Switzerland.

They landed with a jerk. The carabiners automatically unclipped from the lines, leaving their bodies tingling and slumped over. All the blinding white light slowly faded back to color and their strength was renewed. But where had they landed?

They had been expecting mountains and streams and sunshine, but instead they were in some kind of small square room, sitting in front of an empty gray metal table on two cold gray metal chairs. The walls of the room were bare, except for the right wall, which had a door and a large rectangular mirror on it.

Blaine was about to ask his sister if she thought it was a two way mirror, when the door suddenly opened up, and in walked a woman and a man, both dressed in black suits. Their shoes echoed off the floor as they walked over and took seats in the two chairs on the other side of the metal table. The woman, who had long wavy blond hair and striking blue eyes, put a manila folder on the table, folded her hands in front of her, and then smiled at the twins. The man, however, who had jet black hair and eyes to match, just folded his arms in front of him and stared at the children with a frown.

"Hello, Blaine and Tracey," the woman said, in a kind voice. "I am Agent Mac, and this is my partner, Agent . . . Cheese"

Agent Cheese grunted. Agent Mac continued.

"We have a few questions that we want to ask the two of you."

"How do they know who we are?" Tracey wondered worried.

"Yum, mac and cheese sounds delicious" Blaine thought silently.

"We know that the two of you have had contact with Yuroslav Bogdanovich," Mac continued. "What's your affiliation with him?"

Tracey's mind was screaming. "Yuroslav Bogdanovich! He's the crazy scientist that we ran into in Siberia. Who are these two agents? What is this place? They must know about the invisible zip-lines!"

The grumpy male agent suddenly slammed his fists down hard on the metal table and then stood up.

"It's no use, Mac!" he shouted. "You're never going to break these two!"

Agent Mac looked up at her colleague calmly, and then looked back at the twins with a smile. Cheese started pacing the room, while Mac repeated her question.

"Again, let me ask, what is your affiliation with Yuroslav Bogdanovich?"

Tracey didn't know if she and Blaine were really even affiliated with Bogdanovich at all. They had just met him one time, and all they had tried to do then was get away from the crazy man. She trusted this female agent, at least a little, but the male agent was angry and grumpy. She didn't want to tell him anything.

Agent Mac opened up the manila folder and took out several eight by ten black and white photographs and spread them out on the table. The twins were shocked. Every single one of the pictures was a picture of them, on the train in Siberia.

"I'm telling you, Mac!" Agent Cheese shouted as he paced. "You're never going to get them to talk! These two are like locked vaults!"

Suddenly, the door to the room opened up again, and in walked two more black-suited agents, both males with sandy colored hair and shades on.

"What are the two of you doing?" the shorter one asked. "This is our investigation!"

"You got to be kidding me!" Agent Cheese rebutted. "You're pulling jurisdiction now?"

"That's right!" the short sandy-haired agent responded. "Captain Marolf assigned this case to us, and we are going to take the reins right here and now!"

THE SASSAFRAS SCIENCE ADVENTURES

"No way that's happening!" Cheese argued.

"You bet it is!" little sandy shot back.

"Gentlemen, gentlemen," the taller sandy-colored hair agent cut in, with a controlled voice. "Let's not argue in front of the two suspects." The new agent then shot an attempted refined and handsome look Agent Mac's way. "Do you mind if we take this from here, darling?" He asked her.

She rolled her eyes, but then smiled.

"I'm not your darling," she reminded him, as she stood up and started gathering up the photographs she had laid out on the table. "But if Captain Marolf put the two of you on this case, then by all means, proceed."

"Oh c'mon Mac!" Cheese shouted. "You're not really going to let these two butt in are you?"

"Orders are orders," The blond agent responded. "Let's go, Cheese. Let's let these two have at it."

Agent Cheese grunted and then huffed out of the room, followed by Agent Mac, who flashed the twins one more kind smile on her way out. The short agent closed the door behind the two, and then walked over to the table in haste. He took his sunglasses off, as did the other agent.

"I am Agent Pork, and this is my partner, Agent . . . Beans," he greeted, as they both sat down. "We have a few questions that we want to ask the two of you."

"Here we go again," Tracey thought.

"Oh man, pork and beans sounds good too," Blaine thought, rubbing his grumbling tummy.

"We know that the two of you have had interaction with Yuroslav Bogdanovich," Agent Pork informed them. "We want you to tell us what you know about him."

"He's an evil, evil man and we are in no way affiliated with him," Tracey offered.

"He shot at us with his 'Aggrandizer'." Blaine added.

"Aggrandizer?" Agent Pork asked.

"Yes," Blaine confirmed. "It was this machine that could suck up small items, turn them into bigger items, and then shoot them at you."

Agents Pork and Beans shot each other glances as they uttered, "Hmmmm."

"Did he ever say anything about chemical agents?" Pork asked.

The Sassafrases both shook their heads "No". The two agents looked at each other again, as if communicating without words, Agent Beans, the taller agent, then stood up from his chair. He placed both of his palms on the table, leaned forward, and looked deep into the eyes of each twin.

"Okay, Blaine and Tracey, I'm going to shoot you straight."

The twins sighed in relief. They sure didn't want to be shot crooked.

"Triple S had the two of you pegged as possible terrorists."

"What? Terrorists? That's preposterous!" both twins' eyes cried out.

"But I have a polygraph as a sixth sense, and I can tell by looking at the two of you, that you are telling the truth about not being affiliated with Bogdanvich." Agent Beans stood up straight, turned, took a few steps away from the table, and then turned back to face the twins.

"Let's start over." He paused. "I am Special Agent DeBlose, and this is my partner," he said, pointing toward the shorter man, "Special Agent Jorgen Wuthrich. We work for Triple S."

"What's Triple S?" Blaine asked.

"Swiss Secret Service," Agent DeBlose answered.

"We have intelligence that suggests Yuroslav Bogdanovich is currently here in Switzerland, and that he's planning to use chemical agents to poison our country's water sources. When we heard about your encounter with him in Siberia and subsequent survival, we thought you might have been working as his associates."

The Sassafrases shook their heads with an emphatic "No."

"Now that you've been cleared," DeBlose continued, "we want you to help us find this madman and stop him."

Now the twins emphatically nodded their heads "Yes."

"How do you think he is going to try and poison the water?" Blaine asked.

"More than likely, he's going to go for the groundwater," the tall sandy-haired agent said. "Here's how it works: when it rains, a fair amount of water is absorbed into the soil. Just below the soil is a layer of porous rock that can hold water. Then below that is a layer of impermeable rock that water cannot pass through. So the water sits in the porous rock, creating a layer of water known as the aquifer, and the top of this layer is known as the water table."

The Swiss Secret Service agent was clearly giving the twins their SCIDAT data. They had intelligence via LINLOC suggesting this was in fact their local expert, Evan DeBlose.

"The water found in the aquifer is known as groundwater," Evan continued. "And groundwater is a key source of fresh water. As this water soaks

**NAME:** Groundwater
**INFORMATION LEARNED:** The water found in underground aquifers is known as groundwater.

through the rock, it can carve out huge caverns, underground rivers, and waterfalls by dissolving the minerals found in rocks. Rocks such as limestone, for example. When the porous rock layer meets the surface, like on a hillside, a spring can form. Then the groundwater bubbles out, forming a small stream or pool. Spring water is typically very pure and packed with minerals as it has filtered through the porous rock layer.

"Here in Switzerland, the springs, streams, and rivers that flow down from the Alps provide the water for the entire Bern municipality." Agent DeBlose stopped and shook his head, but he looked determined. "We have to stop this maniac, Yuroslav Bogdanovich, from getting to our groundwater with his poison."

Everyone in the room agreed. Evan walked back over to the table and took a seat.

### *I Spy: Waterfall*

"There is a kayaking competition this afternoon out near Reichenbach Falls," the tall handsome special agent declared. "Our sources say this is where Bogdanovich is going to make a trade for the chemicals." "We think that he is going to try and trade the Aggrandizer contraption you two mentioned earlier," Agent Wuthrich said, chiming in. "But what Yuroslav doesn't know is that the person he's trading with is actually one of our agents. We hope to lure him in, nab him, and bring him into custody. What we most need the two of you to do is help us get a positive I.D. on Bogdanovich, since you are the only people we know that have actually seen him face-to-face."

"Don't you have any pictures of him?" Tracey asked.

"We did," Wuthrich answered chiming in. "But strangely, the box of files his pictures were in disappeared from the evidence vault."

The two Swiss Secret Service agents stood up from the table

and walked over toward the door to have a hushed conversation among themselves. Blaine and Tracey leaned toward each other and did the same.

"The invisible zip-lines are supposed to land us as close to our local experts as possible without being detected, right?" Tracey whispered.

"Right," Blaine confirmed. "But at this location, it's like they were expecting us, and they even knew us already!"

Tracey nodded and whispered on. "Do you think Uncle Cecil knows that Evan DeBlose is a Triple S agent? Do you think this fits into his overarching plans for our summer learning?"

Blaine shrugged. "I'm sure it does. Don't you trust Uncle Cecil?"

"Of course I trust Uncle Cecil, but c'mon, Blaine, you know how absent-minded he can be. There are just too many working parts for him to be able to control everything. I mean, at first I thought it was just you, me, and him that knew about the zip-lines, but now it seems like everybody knows. Do you think that's okay?"

Blaine didn't answer because he had zoned out a little.

"Blaine! Do you?" Tracey asked.

"Do I? Do I wh—" Blaine stammered. "Do I think they may have some mac and cheese or pork and beans sitting around that they would let us eat?"

Blaine smiled and answered his own question, "Yes, I do."

Tracey looked at her brother with exasperation and was about to try and get him back on track when agents DeBlose and Wuthrich came back over to the table.

"We have decided to give you two Level Niner clearance," DeBlose declared.

"Level Niner?" Blaine asked. "How many levels are there?"

"I can't answer that, Blaine. It's above Level Niner clearance."

"Oh," the Sassafras boy responded flatly.

"Follow us," Wuthrich commanded. "We have some things to show you."

The twins followed the agents out of the square room into the hallway and were immediately met by a shout.

"You're giving them Level Niner clearance?" It was Agent Cheese. He and Agent Mac had been watching the twins' interview from behind the rectangular mirror, which was indeed a two-way mirror, as Blaine had suspected.

"That is insane!" Cheese yelled. "What if they are lying about not being Bogdanovich's associates?"

"They're not lying," DeBlose replied confidently.

"Oh, how do you know that? Because of your supposed sixth sense? Whatever, DeBlose. I have lied to you before, and you didn't catch me on it."

Instead of responding, Evan and his partner led the twins down the hallway in the opposite direction of Mac and Cheese.

"I'm telling you, DeBlose, this is a bad idea!" Agent Cheese shouted out behind them.

Evan turned and flashed a playful scowl at Cheese.

"We'll see, Zwyssig, we'll see."

Cheese suddenly looked horrified. "Don't use my real name! How dare you use my real name!"

DeBlose just chuckled, turned back, and continued down the hallway. It wasn't a very long hallway, and at the end of it, Agent DeBlose turned to the twins and said in a low voice, "Now that you're Level Niner clearance, I can tell you that those two, Agents Mac and Cheese, are actually the beautiful special agent Adriana Archer and the grumpy special agent, Gottfried Zwyssig."

## Chapter 16: Quick! To Switzerland!

The four walked through a doorway into a large high-ceilinged room that looked like some sort of factory. There were dozens of technicians moving around to and fro in white lab coats. One of them immediately came running up in excitement. He was a skinny little man with a head that looked a tad too big for his body.

"Agent DeBlose! Agent Wuthrich! Welcome! Welcome! And who are these two little tykes with you?"

"This is Blaine and Tracey Sassafras," Evan answered. "Blaine and Tracey, this is Q-Tip. He is Triple S's expert in, shall we say, 'technologizing.'"

Q-Tip nodded excitedly. Then he waved his hand, beckoning for the group to follow him.

"Come over here and see what I've cooked up for you," he said, with abounding enthusiasm.

Q-Tip ran over to a worktable and picked up a pocket knife. "As always, I am sending Swiss Army Knives out with you on this assignment, but I have added an upgrade. The knives still have all their normal utilities, such as flat-blades, serrated blades, corkscrews, etc. They also have some of their past upgrades such as the taser, the flashlight, the telescope, the GPS, the small missile launcher, the toothbrush, the air freshener spray, the glass cutter, and the nunchucks."

The twins shot each other amazed glances. How could this small pocket knife hold all these utilities and crazy upgrades?

"And now, for this assignment, I have added a final upgrade," Q-Tip announced happily. "A parachute!"

Upon this announcement, both DeBlose and Wuthrich had confused looks on their faces. Blaine and Tracey both thought that one of the agents would ask Q-Tip how he fit a parachute inside a Swiss Army Knife, but that is not what they asked.

"Why would we need a parachute on this assignment?"

Wuthrich asked. "We will be in and around the water for the majority of the time."

Q-Tip just smiled.

"You never know when a parachute might come in handy. Next . . ." The technologizing expert moved on as he set the knife back down on the table and skipped over to another area. ". . . we have 'Kayak X!'"

He was now pointing at a very normal looking yellow three-man kayak.

"What does the 'X' stand for?" Blaine inquired.

"Haven't you ever done algebra before?" Q-Tip asked with a smile. "The X can be anything, but it has to be something for sure."

This answer only made Blaine more confused. Maybe that was because he, in fact, had never studied algebra.

"More specifically, this 'X' is: headlights, all-terrain tires, grappling hooks, net launcher, inflatable protective padding, paddles that turn into prop motors, and . . . a taser!"

Now both special agents were smiling.

"I can definitely think of uses for all of those things!" DeBlose said.

"And, last but not least," Q-Tip continued, bounding over to yet one more area, "I have designed thermoses for all of you."

"Wow! What inside of these?" Tracey asked, enthralled.

"Why, hot chocolate, of course," Q-Tip answered, like his answer should have been obvious.

"Oh," Tracey sighed, a little disappointed.

"And . . . " Q-Tip went on, holding up one of the thermoses. ". . . tasers!"

"Impressive." Blaine nodded his head.

## Chapter 16: Quick! To Switzerland!

"You seem to really love tasers," Tracey added.

"That's why they call me Q-Tip," the quirky inventor smiled.

This didn't make any sense to the twins. The skinny man explained. "They call me this not because I have a big head on a small body. They call me Q-Tip because my first invention for Triple S was a q-tip that could be used as a taser."

Now the twins understood. They both hoped they would get the chance to show Q-Tip the taser applications on their smartphones.

Suddenly, the door to the room swung open and in walked a huge barrel-chested man with a gray flattop haircut. Every single person stopped what they were doing and gave the man their full attention. The huge man walked straight over to the Sassafras twins with an intimidating air of authority. He looked the two up and down and then looked at Evan.

"These are the two?" he asked, in the deepest voice the twins had ever heard.

"Yes sir, Captain Marolf, sir," Agent DeBlose answered.

"And who, may I ask, authorized you to give them Level Niner clearance?"

"Well, sir, I just thought—"

"You thought, did you? I don't think you did think because if you did think, you would have thought to ask me what I thought before you thought it was okay to grant them Level Niner clearance!"

"Sir, I am positive they are not working with Bogdanovich," DeBlose said confidently. "We need their help to I.D. and catch this villain before he poisons our country's water."

The wrinkles in Captain Marolf's face eased a bit. "I agree, Agent DeBlose," the big man said. "But next time, you need to pass it by me before you give anyone any kind of clearance. Do you understand?"

"Yes, sir."

Captain Marolf looked at the twins one more time and then back at DeBlose and Ulrich. "Finish briefing these two Niners, and then all of you get out to that kayaking competition. We need to be set up and ready to go—sooner rather than later."

"Yes, sir," the black-suited agents answered in unison.

DeBlose and Wuthrich proceeded to show the twins everything they were cleared to see down in the big lab where Q-Tip worked. They then emerged from the top secret corridors and showed Blaine and Tracey that the Triple S headquarters were actually discreetly and cleverly located inside a huge clock tower called the Zytglogge, right in the heart of the city of Bern.

They also took the time to show the Sassafrases the eleven Renaissance allegorical statues that were located all around the city. Each statue stood on an ornate pedestal over a flowing fountain. These beautiful fountains that were topped with the statues were the pride of the city, but they would soon be spewing poisonous water if they couldn't stop Bogdanovich.

A couple of hours after their tour of the city, Blaine and Tracey found themselves out in a beautiful scenic setting, close to Reichenbach falls. The forest, the rocks, the streams, and the waterfall itself were all truly breathtaking.

A large crowd had shown up for the kayaking competition. There were cars and four-by-fours parked anywhere they could find a spot. There were people that had camped out and people who had come for the day. They were all excited about the start of the competition. It was a dense and loud crowd that would be easy to get lost in. The twins were hoping they would get to go out and enjoy some of the fun, but for now they were sitting in a dark unmarked van, looking at an array of video monitors. With the twins were with Special Agent DeBlose, Special Agent Wuthrich, and the van driver, whose name was Klemens. It was their job to keep their eyes on the monitors for any sighting of Yuroslav Bogdanovich.

The twins' minds drifted back a few days to the first time they had met Bogdanovich on that top secret north bound Siberian train. Yuroslav oozed wickedness when he spoke. His white lab coat and crazed red hair had made him look like an evil version of Uncle Cecil. The pupils in his eyes had actually pulsated when he had interacted with them. And then there was the matter of that contraption he had invented: the Aggrandizer. It could vacuum up small items, turn them into bigger items, and then shoot them.

Bogdanovich had chased them all over the train and shot at them with aggrandized cobwebs, Jell-O, and metal shards. But, thanks to their Russian friend, Sveta Corvette, they had escaped the crazy man. That was, until he buried them in a castle-sized mound of aggrandized snow. The twins had survived that, too, and they hadn't seen Bogdanovich after that. Their hope had sort of been that they would never have to see him again because he was a legitimately scary man. But, here they were now in Switzerland, trying to find him again.

"Okay, everyone, take your places," the deep voice of Captain Marolf sounded out in everyone's earpieces.

Agent Wuthrich slid open the van's side door and stepped out. He was the one who was posing as the chemist who was going to trade the poisonous chemicals to Yuroslav for his Aggrandizer.

"Let's get this guy," Wuthrich declared to his colleagues in the van. He then turned and mingled into the crowd.

Special Agents Archer and Zwyssig had been assigned to watch the perimeter. They were in charge of about a dozen more Triple S agents that were somewhere hidden among all the kayaking fans, dressed in plain clothes, ready to assist in the capture of the crazy Russian scientist.

"This is a beautiful place, is it not?" Evan DeBlose asked the Sassafras twins.

They both nodded as they glanced out of the van's windows, sneaking their eyes away from the monitors for a second.

"Reichenbach waterfall is right in the middle of a deciduous forest," the Triple S agent informed. "Deciduous forests are also known as temperate forests and are characterized by trees that shed their leaves during the fall season. These kinds of forests have four distinct seasons, which are, of course, fall, winter, spring, and summer. Each season lasts about three months. Right now it is summer and it is the absolute best time of year to hold the annual world-famous kayaking competition because the temperatures in the summer average around seventy degrees Fahrenheit. But in the winter, it can get cold out here, as the temperature often drops below freezing. Did you know that deciduous forests typically get between thirty and sixty inches of rainfall per year?"

"Here at Reichenbach Falls and the surrounding areas, that rain plus lots of snowmelt from the Alps, keeps the water flowing through making it a perfect place for kayaking."

"It sounds like you love kayaking," Tracey offered to the secret agent and local expert.

"I sure do," Evan smiled. "I have actually competed in this competition in years past. But now that I work for Triple S, I have to keep my profile a little lower."

"What about the kayak that Q-Tip made for us?" Blaine asked. "Do you think we'll get to use that?"

"I hope so!" DeBlose said, with wide eyes like a giddy schoolboy.

"Are you actually supposed to ride a kayak down the waterfall?" Tracey asked, looking out at the towering Reichenbach waterfall.

"Oh, no!" Agent DeBlose answered. "Riding a kayak down

the Reichenbach would surely be fatal, but there are many other smaller waterfalls that would surely be a blast to ride down. Waterfalls can be just a few feet tall or much much larger. The tallest known waterfall is Angel Falls in Venezuela. It is three thousand two-hundred twelve feet tall!"

"Wow!" both twins stated. "You definitely wouldn't want to kayak down that!"

"No, you wouldn't," Evan confirmed. "Really, a waterfall is just a point in a stream or river where the water suddenly falls downward. Waterfalls typically occur where a stream or river flows from hard rock to softer rock. The flowing water wears down softer rock faster, which creates a drop. Waterfalls can also be formed when the land is changed by a major event, like an earthquake or a landslide."

"How many different kinds of waterfalls are there?" asked Blaine as he and Tracey took pictures of the Reichenbach waterfall.

"There are many different ways to classify waterfalls," Evan answered. "The most common way is by how the water actually falls. For example, Reichenbach Falls would be classified as a horsetail waterfall. This means that the water keeps in contact with the bedrock below as it falls. Another way water falls is in a sheet from a wide river. This kind of waterfall would be classified as a block waterfall. Niagara Falls, in North America, is an example of this. There are als—"

NAME: Waterfall
INFORMATION LEARNED:
A waterfall is just a point in a stream or river where the water suddenly falls downward.

"Look!" Tracey suddenly interrupted her local expert with a scream that sent a chill down her own spine as well as everyone else's. "There he is! There is Yuroslav Bogdanovich!"

## Chapter 17: Tracking down Bogdanovich

### *River Reconnaissance*

Seeing the creepy scientist on the video monitor immediately reminded the Sassafrases how much this man had scared them on the train. He wasn't wearing a white lab coat today. Instead, he was dressed in a sky blue suit. His red hair was still sticking up, but they couldn't tell on the monitor if his pupils were pulsating or not. They did see that he had a big bag slung over his shoulder, presumably with the Aggrandizer inside it. The twins couldn't help but shiver at the sight of the villain.

Special Agent DeBlose sensed the twins' fear, so he battled it with a compliment and a refocusing. "Well done spotting him, Tracey! You too, Blaine! Now let's re-hash the plan."

The twelve-year-olds nodded and settled down.

"Agent Wuthrich will meet Bogdanovich at the vendor cart that sells miniature chocolate kayaks," Evan recounted. "Yuroslav is supposed to give the prompting line: 'Nice day for everyone to have fun in the sun.' Then Agent Wuthrich is supposed to respond by saying, 'Yes, but I'd rather be sharing a cinnamon bun with Attila the Hun.' This will let each know they have found the right person. Jorgan will then give Bogdanovich a drop of the chemical so Yuroslav can test to see that it is, in fact, poisonous."

"Bogdanovich will then hand over the Aggrandizer to Wuthrich, and Wuthrich will give the vial of chemicals to Bogdanovich. We will give Wuthrich a moment to get out of the area, which will give Bogdanovich a moment to think everything is going according to his plan. Then, Agent Archer, plus Zwyssig and the men they are in charge of, will sweep in to take Bogdanovich down and retrieve the vial."

Blaine and Tracey nodded again, gaining more and more confidence.

"Sounds correct to me," Wuthrich affirmed in the earpieces that everyone was wearing. "I think I see him approaching the vendor cart right now. Am I correct in thinking that he is the one wearing the sky blue suit?"

"Yes, that is correct," Tracey responded into the earpieces.

"Okay, then, this deal is about to go down," Agent Jorgan Wuthrich declared.

The Sassafrases and DeBlose watched on one of the monitors as Yuroslav approached Agent Wuthrich near the cart. Both the mad scientist and the special agent got in line like they were going to buy a miniature chocolate kayak.

"Nice day for everyone to have fun in the sun," Bogdanovich stated the prompt.

"Yes, but I'd rather be sharing a cinnamon bun with Attila the Hun," Jorgan came back with the secret response.

Agent Wuthrich then handed Bogdanovich a dropper with a single droplet of the chemical in it. The mad scientist took the dropper and then reached in a side pocket of his bag and pulled out a small container of clear liquid. He took the cap off, squeezed the drop in, and the clear liquid immediately turned green. Yuroslav nodded like he was satisfied. The maniac scientist now reached into the main compartment of his bag and grabbed something that looked like a big water gun.

Blaine and Tracey immediately recognized it as the 'Aggrandizer.' Yuroslav handed the gadget to Jorgan, who quickly put it in a backpack he was carrying. Special Agent Wuthrich then very carefully pulled out the vial of poison and put it in Bogdanovich's waiting outstretched hand. Then, as fast as a lightning strike, the evil scientist shoved the agent to the ground, grabbed the backpack with the Aggrandizer in it, and took off into the crowd. Now Yuroslav Bogdanovich had the poison and the weapon.

Suddenly, with no warning, all the monitors in the van went blank.

"What happened to the video?' DeBlose shouted out. "Wuthrich! Wuthrich! What's happening? We lost our feed!"

Evan tapped at his earpiece like it wasn't working. "Wuthrich! I can't hear you! Are you there? Sounds like we've lost our audio as well. Archer! Zwyssig! Can anyone hear me?"

Blaine and Tracey could hear Agent DeBlose, but only because they were sitting right beside him in the van. Their earpieces were not working anymore either.

"Klemens! What's going on?" Evan screamed, as he made his way toward the front of the van. "Klemens! I said what's going . . . Klemens?"

Klemens, the van driver, was not in his seat in the front of the van. He was gone.

"What is happening?" DeBlose asked out loud. "This assignment is going south real quick!"

THE SASSAFRAS SCIENCE ADVENTURES

The Triple S agent looked at the twins with a far-off-looking stare that rather quickly turned to a look of realization. "We've lost our eyes and ears, and that can only mean one thing."

"What does it mean?" the twins asked.

"It means we have a mole," Evan stated matter-of-factly

"A mole?" Blaine asked. "You mean like the rodent that digs through the dirt?"

"No, not that kind of mole. You know, a mole—a double agent—a turncoat, a traitor."

"Oh, that kind of mole," Blaine mumbled.

"How can you be so sure?" Tracey asked.

"Because all of Triple S's data runs through a private satellite," DeBlose answered. "And to start or stop that data's movement, you must have a password. More specifically, an encrypted password that only high level Triple S agents are privy to. So this malfunction has to have been caused by someone on the inside!"

DeBlose paused again and a grave look of concern covered his face. "One of our agents is working with Bogdanovich! We have to get out there and stop them! If we don't, all of Switzerland's water could soon be poisoned!"

Special Agent DeBlose reached over, and cracked open the van's side door, and jumped out, with the Sassafrases right behind him. The black-suited man quickly took off into the crowd of kayaking fans in the direction of the vendor cart where the deal had gone down. Almost immediately, the three ran into Agent Adriana Archer.

"What happened?" she asked. "What's wrong with our earpieces?"

"The satellite feed must have gotten cut off," Evan answered quickly. "Where is Zwyssig?"

"I don't know," Adriana answered frantically. "I don't know

where any of our men are! But we must do something quickly! Or else we are going to lose the chemicals to that madman!"

"I know, darling, I know." DeBlose shook his head with a hint of playfulness, even in the midst of this stressful situation. "I'll get to the vendor's cart, find Wuthrich, and see what happened. You find Zwyssig and your men and work on getting the satellite connected again."

With that, Evan and the twins took off into the crowd again.

"I'm not your darling," Agent Archer reminded the tall sandy-haired agent as he ran away.

Blaine and Tracey followed DeBlose as he wove in and out of the mass of people. They made it to the miniature chocolate kayak-selling cart in good time. When they arrived, they saw no sign of either Yuroslav Bogdanovich or Special Agent Jorgen Wuthrich. Evan spent only a few seconds looking around before he pulled out his Swiss Army knife and unfolded the telescope.

The telescope, which was one of Q-Tip's many upgrades to the knife, was a tiny little thing, but evidently it worked because after a few moments of scanning the crowd, Evan had something.

"There he is! I've spotted him!"

"Who? Bogdanovich?" Blaine questioned.

"Or Agent Wuthrich?" Tracey asked.

"Bogdanovich," DeBlose replied. "He is heading down to the water. Let me put a mark on him."

"Put a mark on him?" Blaine inquired.

"Yes, a mark," the agent said, as he showed the twins what he was talking about. "The telescope allows you to set a digital mark on any item you can spot through its lens. Then, you can transfer that digital mark to the Swiss Army Knife's GPS. The GPS will then tab the mark if it is stationary or track the mark if it is mobile, allowing you to follow it. So, I just set a mark on Bogdanovich and have already transferred it to the GPS. Now even if we lose sight of him, we can still follow him."

"Cool," both twins said, impressed.

The agent again looked through the telescope and had some unwelcomed additional information.

"Bogdanovich is getting in a speedboat!"

"A speedboat?" Tracey asked, alarmed.

"Yes," Evan answered. "This is not good. This will enable him to get up river quick. He'll be in the area where all the springs and ground waters are in no time. C'mon Sassafrases, let's get Kayak X, and let's get going after him!"

In a matter of minutes, the twins found themselves in the water in the yellow kayak, sitting in two small seats in front of Agent DeBlose, who was sitting in a third seat behind them. The three of them mixed in pretty well with the kayaking enthusiasts around them, except for the fact that Evan was still wearing his black suit.

Evan, Blaine, and Tracey all stuck their paddles in the water and then pushed buttons that caused the blades of the paddles to start spinning. This propeller action pushed Kayak X through the water like it was being run by a powerful motor.

Three backpackers stood at the water's edge and gawked in amazement at the swiftly moving kayak.

"Dude, have you ever seen anyone paddle a kayak that fast?" the first backpacker commented.

"No, man, that is crazy," the second said.

"They've got some sick skills," the third added.

"Dude, wait a second! I think that was those Sassafras twins!"

"You're right, man! That's Blaine and Tracey in that kayak!"

"Those kids are sick!"

Blaine and Tracey could not see Yuroslav or his boat up in front of them, but they knew he was moving somewhere because they could see his mark moving on the Swiss Army Knife's GPS.

"I don't know if these prop motor paddles will move our kayak as fast as Bogdanovich's speed boat is going," DeBlose shouted out as they cut through the water. "But they are propelling us pretty fast, aren't they? The things that Q-tip comes up with never cease to amaze me."

As they went up the river, they passed lots of kayakers. Some were simply paddling down the wide body of flowing water Kayak X was going up, but many others were dropping down into the wide water from smaller flumes on each side. The flumes were cutting through rock and earth in really cool and beautiful ways. Agent DeBlose pointed out some that he thought were especially interesting, and then he started giving scientific information.

"Those smaller shoots of water the kayakers are using are streams," he said. "Streams are bodies of fast moving freshwater. They are much smaller and shallower than rivers. Some of them you can even walk across. Many streams will flow together to form a river. This wider moving body of water that we and our kayak are in right now is a river. Rivers carry water from their source, or beginning towards a larger body of water, such as a lake or ocean. The source of a river is typically found high up in the mountains, where lots of smaller streams come together."

LINLOC  SCIDAT

**NAME:** River
**INFORMATION LEARNED:** Rivers carry water from their source, or beginning towards a larger body of water, such as a lake or ocean.

Tracey, who was in the front seat, oohed, and Blaine who was in the middle seat, aahed, as they saw one kayak whip down a zigzagging stream to join the river. Then another kayak shot off a waterfall to splash down into the river.

"All rivers have three key stages," the Triple S agent continued. "The upper stage is where the river's course is steep and the water moves very fast. This is where the many streams are joining together to create the river. The water in the upper stage is very clear. The middle stage is where the bed of the river smooths out and the path of the river begins to flatten. The river widens in this stage and makes wide loops known as meanders. Smaller rivers may also join at this point to create a larger river. Then, the lower stage is where the river flattens out to join the lake or ocean. The water in this stage is packed with sediment so it appears very muddy. At this point, the river may also split into channels to form what is known as a delta."

The three used the prop motor paddles to guide Kayak X up and around a bend in the fast moving river they were in.

"Rivers not only carry water," DeBlose continued, "they also pick up loose sand, soil, and rocks. They carry this material along, which causes the bed, or bottom, of the river to become wider and deeper."

Suddenly all three Kayak X-ers saw something that made their hearts leap. Up ahead of them in the river was a speedboat! It was a speedboat with the mad scientist in it! They had caught up with Bogdanovich! The boat was still moving, but not very fast. Yuroslav may not have even been aware he was being followed. However, when he turned his head back and saw the three in the approaching kayak, he was aware.

He cranked up the throttle and jetted up the river. With the three prop motor paddles still in the water, the Triple S agent and the double Sassafras kids continued their pursuit. Now it was easy to tell his motorboat was faster than their kayak, but before they lost sight of him again, they caught a glimpse of something looming up ahead that would surely aid them in capturing him—a huge towering dam.

The presence of the dam created a dead end in the river. There was now nowhere for Yuroslav Bogdanovich to go. Hopefully, now they would be able to capture him, confiscate the Aggrandizer, and get back the vial of poison before he was able to use it to taint Switzerland's water sources.

The twins knew this would be easier said than done because Bogdanovich, after all, was a madman. Yuroslav sped up to the base of the dam, steered his boat in a wide loop, and then started speeding directly towards the kayak.

"He's going to crash right into us!" Tracey shouted out.

"Not a chance," Agent DeBlose responded. "Blaine and Tracey, pull out your Swiss Army Knives and fire a few small missiles at him! They won't blow him up, but they will daze him enough to make him think twice about coming in for a smash."

"Oh, yeah," the twins thought. "We have small missile launchers on those pocket knives Q-Tip gave us!"

Each Sassafras pulled out the packed utility knives and took aim at the fast-approaching speedboat. Neither of them was really sure how the small missile launcher even worked, but they both fired.

Tracey's small missile shot out with a scream and some sparks and zipped toward Yuroslav like a bottle rocket, but instead of dazing the madman, it just glanced harmlessly off the side of his boat. Blaine's small missile didn't fire out at all because he had accidentally pushed the air freshener spray instead. It smelled delightful, like rose blossoms and lavender, but it did nothing to stop the speeding boat that was barreling toward them.

"Let's try the net launcher!" Evan shouted out. "Maybe we can stop him by tangling him up in that!"

Yuroslav Bogdanovich was now close enough that the twins could see his pulsating eyes and hear his wicked laugh. Surely he was about to pummel them.

Agent DeBlose pulled a small lever on their boat marked "NET" and immediately, a big wide net shot out of the front of Kayak X. DeBlose had aimed it perfectly, and it was sprawling directly toward the terrorist, but at the last second Bogdanovich cranked on the speedboat's steering wheel, pulling the vessel hard to the right, allowing it to elude the launched net. The net fell meaninglessly into the river, Kayak, X nearly tipped over because of the waves of the speedboat's wake.

The three watched as Yuroslav guided his boat over to the river's edge, where they all saw something they hadn't seen before. There was an agent in a black suit standing at the edge of the forest, next to a jeep that was parked on a small dirt road.

### *Lakeside Lasso*

Blaine's and Tracey's first thoughts were that this was Agent Wuthrich, here to help them. But why would Agent Wuthrich be wearing sunglasses and a wide-brimmed hat to disguise himself? When Bogdanovich got to the shore, the agent actually helped the scientist out of the boat and into the jeep.

"It's the mole," Evan DeBlose shouted. "I can't tell who it is because of the hat and glasses, but that is the turncoat agent that is helping Bogdanovich! C'mon, Sassafrases! Let's get over there and get them!"

Blaine and Tracey were all for capturing the scientist and the mole, but how could they chase a jeep? They were in a kayak now, and when they hit land, they would be on foot. The jeep sped off up the dirt road into the deciduous forest, well before DeBlose and the Sassafrases got to the shore. There was no way they were going to catch the two now. Or was there?

The moment Kayak X hit land, Evan pulled another lever that was marked 'TIRES,' and immediately four all-terrain tires folded out from the bottom of the kayak.

"Tracey! Step on the gas!" Evan implored.

"Huh?" the Sassafras girl asked.

"Under your right foot there is a gas pedal," the Triple S agent said. "Step on it!"

Tracey stamped down hard with her right foot. When she did, she felt the pedal. When she felt the pedal, Kayak X took off up the dirt road like an ATV.

"Now stick your paddle in that hole right in front of you. Then use it as a steering wheel," Evan added.

Tracey quickly obliged, and she now found herself driving the amazing vehicle. Though they were currently more than a little anxious, both twins had to smile right now. Kayak X was awesome!

The Sassafras girl drove the yellow craft up the narrow dirt road with skill. It was a pretty steep road that looked like it was heading up toward the top of the dam. Blaine was very proud of his twin sister. She was doing a great job driving, which was especially impressive considering the fact that the only things she had ever driven before this were bumper cars and go carts.

They had lost sight of the jeep, but special agent DeBlose still had Bogdanovich's mark being tracked on the GPS. After about ten minutes of dusty driving, they saw that the road did indeed lead to the top of the dam. It led out of the forest and onto a paved road that spanned the entire distance of the gargantuan water-stopping structure.

Tracey eased the pressure she was putting on the gas pedal, and Kayak X rolled slowly from the dirt road up onto the dam's paved road. Now, on one side there was a huge lake and on the other side there was a rushing river. There were steep drop-offs with no guardrail on either side. The drop-off to the lake looked long. The drop-off to the river looked eternal.

Tracey was thinking that maybe her first driving experience should be coming to an end. At the other end of the dam sat an empty jeep. It was parked next to a large concrete building that was attached to the dam itself. From this vantage point, the three could only see one door that led into the building. And that door was standing open.

"An empty jeep, an open door; it's almost like they are daring us to come in after them," Special Agent DeBlose mused.

He then addressed his driver. "Tracey, are you ready to do something brave?"

The Sassafras girl nodded her head but didn't answer.

"I need you to drive us off the dam and into the lake."

"What?" Tracey asked alarmed. "Off the dam and into the lake?"

"Exactly," Evan answered. "That building at the other end of the dam is a hydroelectric power plant. Bogdanovich and the turncoat Triple S agent are trying to lure us into the front door, but I think it's safe to bet there is another entrance on the lake side of the plant. If we enter in through that door instead, I am hoping our presence will go unnoticed. So what I need you to do, Tracey, is drive Kayak X off this road and into the lake."

Blaine put his hand on his sister's shoulder. "You can do it, Trace," he encouraged, with full confidence in her.

With no hesitation, Tracey slammed on the gas and used the paddle to steer the yellow kayak right off the road. Kayak X and its three passengers shot out off the edge of the dam and then dove down nose first toward the lake. They soared silently, with baited breath. Then, all at once, they hit the water with a jolting splash. Blaine and Tracey each gulped in gasps of air before the kayak went completely underwater.

"Oh, no," the twins thought. "What have we done? Maybe dropping down from the dam into the lake wasn't such a good idea." But before the twelve-year-olds could spend too much time second-guessing, the yellow kayak popped back up out of the water.

"Well done, Tracey! Well done indeed," Evan complimented.

The three kayakers, now used their paddles like normal paddles and made their way slowly and curiously across the surface of the lake toward the hydroelectric power plant building.

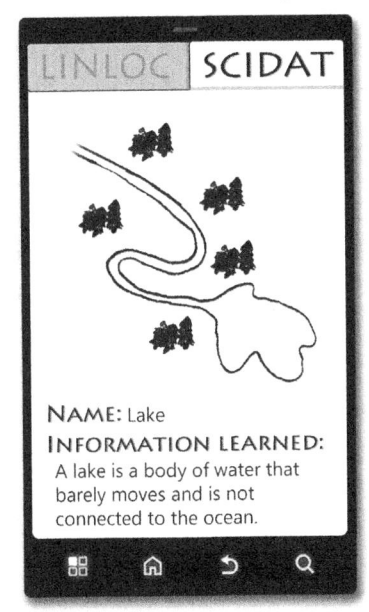

"A lake is a body of water that barely moves and is not connected to the ocean," Special Agent DeBlose said, as he looked out across the beautiful lake they were currently on. "In other words, a lake is basically land-locked. Water flows into lakes from streams and rivers. Lakes can form as a river finds its way into a basin and fills the area with water before continuing its journey downward. Lakes can also form when a bend in a river narrows and is eventually cut off by the deposit of mud and rocks that have been carried in by the river. Another way a lake can be formed is by man."

"By man?" the twins asked.

"Yes," DeBlose answered. "For instance, the lake we are on right now was formed because of this manmade dam. There are also lakes that are dug out by humans and their machinery. These kinds of lakes are known as reservoirs."

The Saassafrases snapped pictures and nodded in understanding.

"Lakes are typically filled with fresh water," Evan continued. "But there are several saltwater lakes. Lakes become salty when there is no outlet for the water. Instead of flowing out, the water evaporates, leaving behind minerals that make the remaining water salty. All lakes, including freshwater lakes, lose water due to evaporation, so they need to be continually fed by water or they will dry up."

As the Triple S agent finished giving scientific information about lakes, the kayak silently glided up to the lakeside of the power plant, and as Evan had suspected, there was a door on this side. Not only was there a door, but there was a metal ramp leading up to that door from the water.

They steered their kayak over to the ramp, and as soon as the all-terrain tires, which were still unfolded, hit metal, Tracey pushed on the gas pedal once again and drove Kayak X up the ramp to the door. The Triple S agent and the twins got out of their boat, and Evan quickly but carefully grabbed at the door handle. It was locked. He pulled out the toothpick and the tweezers from his Swiss Army Knife and picked the lock in five seconds flat. The Sassafrases were impressed.

The three quickly ducked into the door and silently closed it behind them. They now found themselves in a dark room that had large pipes and ductwork running all over the place, making the space feel cramped. The sound in the room was whining and rumbling so loud it almost felt like it was a physically present object. The only light was coming from sporadic dim red bulbs. It would surely be much easier to hide in here than to find someone who was hiding.

Blaine and Tracey followed Special Agent DeBlose as he stealthily made his way from the entry room into a connecting corridor. They also held their Swiss Army knives in front of them like he was doing, ready to tase anybody that jumped out at them. It was a large building, full of loud, dark, winding corridors and hallways. It was like walking through a maze except they weren't trying to go from start to finish. They were trying to go from start to Bogdanovich.

"How are we ever going to find the madman in this eerie place?" the twins thought.

They carefully made their way around a sharp left turn and then stepped through a metal doorway. That began the start of a long straight corridor. Agent DeBlose suddenly stopped. He pointed two fingers at his eyes and then turned those fingers and pointed down the corridor. The twins followed the two fingers and saw what the Triple S agent was pointing at.

There, at the other end of the hallway, illuminated in dim red light, was Yuroslav Bogdanovich. He was looking in the opposite direction, and it didn't seem like he had noticed their presence.

"Let's go," DeBlose mouthed silently.

He then swiftly yet soundlessly began making his way down the long corridor. Evidently, he thought this was a great chance, or maybe the only chance, to catch the crazy scientist by surprise. The twins weren't exactly sure how they were going to aid the special agent, but they followed right behind him, nonetheless. They were approaching Bogdanovich fast, and he still hadn't looked their way. Their chance for success seemed to be soaring.

All at once, a black-suited shadow shot out from a dark nook and tackled Agent DeBlose to the ground. Blaine and Tracey were going too fast to stop, and they toppled right over the top of Evan and his attacker. Not ones to stay down vey long, the Sassafrases both jumped to their feet, ready to tase with their Swiss Army Knives. But before they even knew what happened, the knives were knocked out of both of their hands by a tiny set of nunchucks.

Evan would have gotten up to help them, but he had a foot on his chest holding him down. Now the twins could see the face of the attacker in the dim red light. It was Agent Jorgen Wuthrich.

"Sorry," he apologized.

The short yet strong agent then reached down and tased his

partner, followed by Blaine and Tracey. All three were immediately zapped unconscious.

The Sassafras twins slowly blinked their eyes open, and saw that they were not inside of the hydroelectric power plant building anymore. The first thing they made out, besides the sunshine, was a big yellow banana. No, wait, it wasn't a banana. That was . . . a kayak...Kayak X. It was on the ground in front of them. Also in front of them was Yuroslav Bogdanovich. Next to him was . . . Agent Zwyssig.

Why was Agent Zwyssig here? The twins were still a little groggy from being tased; were they seeing this right? Also, their hands were tied behind their backs—why were their hands tied up? Blaine and Tracey somehow managed to stand up on their feet next to each other. Blaine looked to his left, and already standing there tied up and facing Bogdanovich and Zwyssig was Special Agent DeBlose. Tracey looked to her right, and standing next to her, also tied up and facing the two, was Agent Wuthrich.

What was happening? Both twins glanced over their shoulders and saw that just inches behind them was the huge drop-off to the river. They were on top of the dam.

"So glad you finally decided to join us, children," Bogdanovich sneered in a smug voice that immediately flooded the twins' minds with memories of their run-ins with him on the train in Siberia.

They now realized how dire their situation was. Yuroslav reached into the bag that was slung over his shoulder and pulled out two items. The vial of poison and the Aggrandizer.

"Vith the help of a mole vorking inside the Triple S, I vas able to acquire this!" the wicked scientist said as he held up the vial.

"Vithout giving this avay," he now held up his Aggrandizer.

"Zwyssig is a mole!" Blaine blurted out.

"No, Wuthrich is a mole," Evan said, staring with disdain down the line at his partner .

"I'm not a mole!" Wuthrich defended himself.

"Then why did you tase us?" DeBlose asked.

"Because Bogdanovich said if I didn't tase you, he would end her life!" Jorgen now pointed a little ways down the road.

Everyone turned to look and now saw that Agent Adriana Archer was also present here on the dam, gagged and tied up.

"What?" Tracey questioned. "How did she get here? How did Wuthrich and Zwyssig get here? How did the kayak get here? I thought it was behind the—"

"Never mind that! Stop your squabbling!" Yuroslav shouted. It was obvious he wanted to be the one doing the talking.

"Who the mole vas doesn't really matter," the crazy red-headed scientist uttered. "Vhat really matters is that you recognize my brilliance before I end all of your lives!"

"You're not brilliant," Agent DeBlose retorted defiantly. "You're insane, and you're a coward!" The tall sandy-headed agent then looked at Zwyssig, who still stood at Yuroslav's side. "And you," he added in disgust. "You're even more of a coward, you turncoat! How can you choose to stand with his madman? How can you so easily betray your people and your country?"

"Enough!" Bogdanovich screamed with his pupils pulsating.

He stuck the nozzle of the Aggrandizer down to the ground and vacuumed up some pebbles. The machine made its strange noise as the pebbles enlarged in the internal capacitator. Bogdanovich aimed the contraption at Evan with hate in his eyes.

But before any aggrandized pebbles flew, Agent DeBlose swung his leg up and kicked the Aggrandizer out of the scientist's hands. The big plastic gun flew up into the air, came down, bounced once on the edge of the dam, and fell the long distance down to the river. Agent Evan DeBlose looked proud of himself.

Yuroslav, however, was livid. His pupils pulsated even faster than before, and his uncombed red hair seemingly began to move on its own. The crazy scientist normally displayed an evil confidence, but right now, he was only displaying evil anger. He made a motion to Zwyssig, who immediately grabbed Evan and shoved him into the kayak. The turncoat Triple S agent then started to push the yellow boat over the edge of the dam.

"Them too," Bogdanovich shouted, pointing at the twins.

Zwyssig obeyed, but as he turned his face away from Yuroslav, he winked at the twins and flashed a reassuring smile.

"It's going to be okay," he whispered. "Inflate, then grapple."

The jet-black-haired agent then thrust Blaine and Tracey down into the kayak next to DeBlose, but as he did, he somehow managed to cut the binds that were tying their hands and return their knives to them. Gottfried then gave the yellow boat a tremendous push. Up and over the side of the dam it went

The three were now falling fast. Yes, the kayak was designed to land and float in water, but a fall from this height, even with a water landing, would surely be fatal.

"Blaine and Tracey, are your hands free?" Evan DeBlose shout-questioned.

The Sassafrases nodded and held on.

"Then by all means," the agent implored, "inflate Kayak X's protective padding!"

Both twins immediately remembered one of Q-Tip's additions to this kayak was inflatable protective padding. As they plummeted downward, each twin looked around frantically for any kind of button, lever, or knob that would activate the padding.

Blaine was frantically talking out loud as they dropped, "Tires—no, don't need those right now. Net launcher—no.

Grappling hook—hey, that's cool! But that's not what we're looking for right now. Oh no! We're only yards away from slamming into the river! Where is the...there it is—protective padding!"

Blaine slapped down hard on the button he had spotted. Immediately, a round see-through layer of inflated plastic exploded out around Kayak X. Less than a second after this happened, Evan, Blaine, Tracey, and their now ball-shaped kayak slammed onto the surface of the river.

But they did not sink. They did not get smashed. Instead, they actually bounced. They bounced off the river, spun in the air, hit the dam, bounced off it, and hit the surface of the river again. They bounced one more time before finally coming to a rest, floating safely where the river and the dam met.

The twins felt dizzy. Agent DeBlose, however, looked determined.

"Splendid job, Blaine!" he congratulated. "Now, cut the ropes off my wrist and then let's get back up there!"

"How are we going to get back up there . . ." Tracey started to ask. She paused and then answered her own question. "Oh, the grappling hook!"

Evan nodded as Blaine used his knife to free the agent's hands. DeBlose quickly hit the protective padding button again, which deflated the padding and pulled it back inside Kayak X. Then, he flicked the grappling hook knob. A three pronged grappling hook, with rope attached, powerfully shot up out of the kayak's nose and rocketed directly toward the lip of the tall dam. In just moments, it did its job and securely hooked its prongs somewhere on top of the dam.

"Hold on!" Secret Agent DeBlose urged as a mechanical noise now sounded out from somewhere inside Kayak X.

The rope started recoiling back into the kayak and began pulling the boat up the face of the dam like a winch. The yellow kayak didn't rub or scrape against the concrete surface of the dam because its tires were still out. In a way, they were actually driving, completely vertically, back up to the top. It was fun, but it was also more than a little difficult to hold on.

Up, up, up they went at a fairly rapid speed. The question now was what awaited them at the top. They would soon find out! The winching grappling hook pulled them over the brim, from the face of the dam to its top, and what the three saw now seemed similar to the scene they had fallen away from, but in reality it was altogether different.

All the same people were there, plus one, but in different positions. Agents Wuthrich and Zwyssig were now both gagged and tied up and laying on the road. Agent Adriana Archer was unbound and gag-free, and she was currently climbing into a helicopter that was being piloted by Klemens!

Yuroslav Bogdanovich was already sitting in the helicopter. When he saw that the three in the kayak had returned, his eyes filled with surprise, but he quickly recovered, his arrogance returning.

"You're too late!" he shouted out. "You may have destroyed the Aggrandizer, but I still have the vial!"

The wicked scientist now started laughing. Adriana, who'd originally had her back to the three, now turned to face them. She began laughing, as well.

"Adriana? Darling?" Evan questioned, vulnerable in his betrayal. "It was you? You are the mole?"

'I'm not your darling," the beautiful blonde responded with hubris that matched Yuroslav's. "And, yes, I am the mole. You thought it was Jorgen or Gottfried that had turned, but everything they did today, they did thinking they were helping to keep me alive. All along it was me. I am the mole. I have been working with Yuroslav for years!"

"Adriana," DeBlose pleaded. "Please don't do this. Please

don't poison the waters that feed Switzerland."

"Don't worry. That is not our plan," Archer laughed. "We have something much bigger in mind than that."

Evan looked relieved and confused at the same time.

"We are going to take this vial back to Siberia," Adriana continued. "Now that Yuroslav controls all the secret underground labs there, this poison will prove to be very useful toward our purpose."

"Yes! Yes!" Bogdanovich now shouted, taking over. "I am the greatest scientist this vorld had ever known. All of Siberia vill one day be mine!"

"Not if I have anything to do with it!" Special Agent Evan DeBlose shouted as he all at once rushed toward the helicopter, but Klemens was quick at the controls.

He quickly picked the aircraft up off the ground and then angled the spinning propellers down toward the hero Triple S agent. DeBlose jumped back untouched, but Klemens wasn't done. He now guided the helicopter forward with rotating blades still angled down.

Evan continued to backtrack toward the spot where the twins were, over at the edge of the dam. Before any of the three knew what was happening, the helicopter full of moles and terrorists had chased them off the dam. They were all falling again, down toward the river far below. And they weren't tucked away in Kayak X this time.

THE SASSAFRAS SCIENCE ADVENTURES

## Chapter 18: The End of Earth Science
### *Bonus Data*

They could almost hear the skinny little technology expert saying it now. "You never know when a parachute might come in handy."

Q-Tip had definitely been right about that. Blaine and Tracey Sassafras were floating safely down to the river's edge from the top of the dam with the help of their Swiss Army Knife parachutes.

Triple S agent, Evan DeBlose, had already landed unhurt on the shore below. The twins soon joined him, landing in a similar fashion to how they did after invisible zip-line travel, but they kept their sight and hearing this time.

The three folded up their parachutes as they all looked at each other with relieved yet unsatisfied looks. The beautiful waters of Switzerland would not be turned to poison today, but Yuroslav Bogdanovich had escaped. And there was no telling what that madman was determined to do.

Back at Triple S headquarters inside the Zytglogge in the small square room with the two-way mirror, the Sassafras twins were careful not to taze themselves as they sipped delicious hot cocoa from their Q-Tip designed thermos. Captain Marolf was just about finished debriefing them. The big barrel-chested man with the gray flattop had obviously been disappointed that two of his agents had been turned by Bogdanovich. Marolf had said there was already an operation in progress to hunt down the mad scientist and the two turncoats (Agents Archer and Klemens). They hoped to retrieve the vial of poison before the three could use it in Siberia or anywhere else.

The twins were not surprised when the Captain had told them that Evan DeBlose was the Special Agent in charge of this operation. The big man now stood up from his chair and reached out to shake the twins' hands. Blaine shook Marolf's hand first and almost cried because the captain's grip was so strong. Tracey, however, held it together much better than her brother.

"Thank you, Blaine and Tracey," Marolf said. "The two of you have served the country of Switzerland well. Don't be surprised if we contact you again for future missions."

The Sassafras twins nodded as the big captain left the room, leaving them alone in the exact spots they had originally landed in at this location. Tracey took another sip of cocoa as Blaine massaged his sore hand. Then, the twins took the time to enter all of their SCIDAT data. When they finished, they sent it in, along with the pictures. Next, they opened up the LINLOC application.

"The Left-Handed Turtle?" both twins questioned out loud as they read the next location. There were coordinates: longitude

-80° 20' 15", latitude 37° 22' 17", but there were no topics for study.

"Are we finished with earth science?" Blaine asked.

"I guess so," Tracey answered. "We usually zip back to Uncle Cecil's basement, but I guess this time we are zipping back to Uncle Cecil's neighborhood market."

Both Sassafrases chuckled.

"There's no telling what that crazy man has planned for us." Blaine grinned with obvious affection for his eccentric uncle.

The twelve-year-olds put on their helmets, strapped on their harnesses, calibrated the three-ringed carabiners, and then zipped off through places at the speed of light. They smiled as the light swirled around them. Surely this was the best kind of traveling anyone had ever experienced. It was also always sweet to be zipping back home after completing a leg of their science-filled adventure.

They landed with a jerk. Their carabiners automatically unclipped from the invisible lines. They slumped down and waited for the white light to fade into normal color and their strength to return.

When normalcy had returned, they found they had landed in a . . . dumpster? It was a dark, metallic, and full of newspapers. This wasn't the Left-Handed Turtle!

The twins wondered, as they often did when their landing spots were strange or unexpected, if they had calibrated their carabiners incorrectly. That question in their minds was shot to oblivion when they suddenly heard a voice from inside the dumpster with them.

"Train! Blaisey! You made it! You did it! You successfully completed your earth science leg! Well done!"

"Uncle Cecil?" Tracey asked. "You're in here too?"

"I sure am!" the redheaded scientist said, popping up from under a mound of newspapers. "And not just me, but President Lincoln too!"

The friendly prairie dog popped up from the paper on cue.

"But LINLOC said we were supposed to land at the Left-Handed Turtle," Blaine said. "Not in a dumpster."

"Well, actually, it's not a dumpster," Cecil explained. "It's a recycling bin, and it's in the parking lot of the Left-Handed Turtle."

"Ohhh," the twins replied in unison.

"First there was zoology," Cecil waved one arm slowly in front of him, speaking as if he was talking about something far out in the universe. "And then there was anatomy. Next came botany, and then earth science. And the amazeriffic science learning twins, Train and Blaisey Sassafras conquered them all by grasping the knowledge that was offered and sticking it in their brains and their phones!"

Cecil Sassafras smiled a huge smile and stuck both his arms out wide. "I am so proud of you two! You just keep clickity-clacking along and learning more and more and more! You two are unstoppable. You're fantabulous! You're intelligenius! You're dupersupersasstasticful! My mind is blown!"

The twins' minds were blown as well. They were amazed that their uncle could retain so much energy. They were amazed that he could keep making up such crazy words. And they were amazed that he could continue to facilitate their ability to fly around the globe on invisible zip-lines to learn science face to face.

They were having such a fantastic time, even if they were currently sitting in a big metal box out in a parking lot.

"So, Uncle Cecil, what's the deal with the recycling bin? Why did you have us land here instead of in your basement?" Blaine quipped.

"I'm glad you asked, Train," Cecil answered. "It's because of your bonus data!"

"Our bonus data?" The twins asked.

They had nearly forgotten that every time they made it back after successfully completing all that was required of them for a given subject, they would receive a special text on their phones entitled "Bonus Data." It was like a miniature gift from Uncle Cecil and President Lincoln for all their success and hard work.

Blaine and Tracey pulled out their phones and, sure enough, they had both received the textual gift. Blaine immediately started reading the text aloud.

### BONUS DATA

As humans, our technological advances can hurt our planet. For instance, our need for paper requires cutting down trees. Our need for food and other goods has changed the earth's landscape through farming and mining. Our need for electricity and mass produced products has created pollution and trash that is damaging to our environment.

To help protect our natural

Blaine paused his reading, and Tracey picked up where he ended.

> **BONUS DATA**
>
> As humans, our technological advances can hurt our planet. For instance, our need for paper requires cutting down trees. Our need for food and other goods has changed the earth's landscape through farming and mining. Our need for electricity and mass produced products has created pollution and trash that is damaging to our environment.
> To help protect our natural

Tracey looked up from her phone as she stopped reading.

Cecil smiled, impressed, and then the redheaded scientist shouted in elation with his head up and his arms outstretched. "That's some bonus data jam packed full of great ideas, is it not? The recycling bin we're in at this moment is a product of planet-loving thinkers and their ideas. If you put a bin in an easy to access, high-traffic area, a lot of people will take the opportunity to recycle their paper, plastic, and other used goods!" When their uncle was finished, he put his head and hands down.

"This bin also has a second purpose," he said, in a cautious and much quieter voice as his facial expression changed and his shoulders slumped. "You wanna know what that is?"

Blaine and Tracey nodded, suddenly full of curiosity.

"This recycling bin is also an anti-dog tank."

The twins' curiosity was now layered.

"An anti-dog tank? What does that mean?" Blaine asked.

"President Lincoln and I have added wheels and a motor to this metal bin, giving us the ability to drive it back and forth from our house to the supermarket without being vulnerable to an attack by Old Man Gusher's dog. That guardian beast will prey on us no longer!"

Blaine and Tracey did all they could to hold in their laughter. They weren't sure why Uncle Cecil was so scared of the guardian beast, which was actually only a miniature poodle. But they didn't want to make their uncle feel any worse about the dog than he already did.

"Okay, then!" Tracey exclaimed. "Let's ride this anti-dog tank back to 1104 North Pecan Street!"

Cecil now switched back from scared to excited and proceeded to open up a small panel box on the back wall of the bin. He cranked an ignition switch and the makeshift tank rumbled to life.

"Just so the two of you know, this motor runs on used French fry oil and Brussels sprouts," Cecil shared as he moved from the back of the recycling bin to the front, where there was a small steering wheel the twins hadn't noticed yet.

There was also a long narrow rectangular slot above the steering wheel that allowed all four of the anti-dog riders to peek out all at the same time. The eldest Sassafras guided the recycling bin out of the Left-Handed Turtle parking lot and onto the sidewalk that led home. They drove cautiously, slowly passing different houses in the neighborhood. Their eyes remained steadfastly peeled for any canine advances.

When Cecil made the turn onto Pecan Street, he brought the bin down to an even lower speed. This was the danger zone. Cla-clunk, cla-clunk went the tank as they slowly rolled over the gaps in the sidewalk. They passed 1111 on the right and 1112 on the left. There was no sign of the guardian beast. Cla-clunk, cla-clunk, 1109 and 1110 were now passing by. No barks or growls could be heard.

And then, there it was—1107 North Pecan Street. The dwelling of the beast. They saw the porch from which he kept watch, but still there was no sign of the poodle.

"Isn't that just the way it would work," Cecil whispered, managing a chuckle. "On the maiden voyage of the anti-dog tank, we aren't even going to spot the dog we are anti-dogging."

Blaine and Tracey chuckled with their uncle. That was, until their eyes spotted something through the rectangular slot that sent shockwaves to their hearts. There, wandering aimlessly down the sidewalk, with frizzy blond hair and a blank look in her eyes, was Summer T. Beach!

The Forget-O-Nator had worked perfectly. It had done exactly what it was created to do when it had wiped a brain completely clean of memories. He had experienced his biggest success to date when he had abducted Summer. He'd gotten her inside the Forget-O-Nator and had washed away the contents of her mind into a sea of nothingness. But if all this was in fact the case, then why was he still so unsatisfied? Why did he sit here in his basement unfulfilled?

After forget-o-nating Summer, he was pretty sure he had actually experienced joy as he watched her stumble around with no memories, no knowledge, no anything in her brain. She was still herself, but it was almost like she was a zombie version of herself. The frizzy-haired scientist didn't even remember her own name.

He had briefly felt like his quest for revenge on Cecil Sassafras had been completed. After all, he had ruined the mind of the one that loved Cecil most. Now Summer didn't even know who Cecil was. Additionally, she could no longer help those twins with their science learning.

But after only a few moments of feeling accomplished, a deep flood of discontentment had consumed his dark heart. So much so that he had let Summer go. Seeing her brainless wasn't giving him satisfaction. She was now utterly useless to him or anybody else. So he had just let her walk up the stairs and out the door.

He sat here now with a frown on his face, trying to stay afloat in these waves of nonfulfillment.

"What have I really accomplished," he thought. "Sure, I destroyed Summer, but really, that is just a miniscule part of the big picture. She is just one of the local experts. I haven't stopped anything. Those twins will still be able to zip around the globe on the zip-lines learning science, and therefore Cecil Sassafras will still be happy. Revenge has not been accomplished! Vengeance is not mine!"

His fists clenched and his hairless brow furrowed angrily. He stood up and began an indignant pace around his basement.

"I have to continue on in my quest! I have to stop those twins! I have to crush Cecil Sassafras's dreams!" he shouted, now talking out loud to no one but himself.

"Maybe that is the problem," he paused. "Maybe I have been going about this all wrong. Maybe all I need is the help of someone else. Cecil has President Lincoln. Summer has Ulysses S. Grant. Blaine and Tracey have each other. Maybe I need a partner or an evil sidekick. Maybe if I had this kind of help, I could accomplish all the revenge I long for."

He paused again and smiled a toothy wicked smile. An

entirely new strategy had just entered his mind. "Those twins just finished another scientific subject, and I'm sure in the next day or two they will be starting their next subject. What if I put some real time and effort into recruiting an evil partner? What if I zipped back to past locations, abducted some of the worst villains from their journey, and put them in the Forget-O-Nator? Then in their brainless state, I can mold them into who I want them to be. For if their minds were absent of memory, then I could tell them to do whatever I wanted, and they would have to do it, right?"

The Man with No Eyebrows's angry pace had now stopped, and he was actually getting excited. "Yes! This is what I am going to do! And I'm going to start right now!"

***Jumbled Geology***

Mrs. Pascapali had spotted Summer. The nice old widow

who lived at 1106 had hobbled down from her front porch to see if she could help the poor young woman.

"What happened to you, sweetie?" the widow asked compassionately.

"I...don't know," Summer answered.

Across the street, Cecil, Blaine, and Tracey had all jumped out of the recycling bin, and started running to where the two woman were standing. They arrived in time to hear Summer answer Mrs. Pascapali's question.

Blaine and Tracey were in a state of shock. Not only was it weird to see Summer here on Pecan Street, but it was even more weird to see her so blank. Why was she acting like this? How in the world had she gotten here? The last time the twins had seen her, it had been in Alaska out in the field playing paintball. At that time, she had been her normal self, talking with excitement and enthusiasm and doing Summer Beach type things like happy jump-dance-hugs.

This lady that stood in front of them right now looked like Summer Beach, but the twins were not convinced it really was Summer. How could it be? This woman had lifeless eyes and no trace of joy.

"Do you need a place to stay? Do you need something to eat?" Mrs. Pascapali asked, putting a gentle hand on Summer's shoulder.

"I don't know . . . I just don't know . . . I'm not . . . I can't . . . I don't remember anything." Summer began to cry.

"There, there, sweetie. It's going to be okay. Why don't you come on inside. I've already got a pot roast in the crockpot, and I have a nice comfortable guest room if you decide you need to stay. Okay?"

Summer sniffled, nodded, and wiped some of the tears from her eyes. Mrs. Pascapali took her hand and began leading her into the house. Blaine and Tracey were speechless. Uncle Cecil was at a loss for words, too, but he did manage to offer his assistance if Mrs. Pascapali needed any extra help taking care of Summer.

"I think we'll manage just fine," the widow answered. "But if we need anything, I will call you, Cecil."

A bit later, down in the cluttered and messy basement, Cecil attempted to give the twins an exciting introduction to the subject that they would be studying next, but it was obvious the scientist's heart was troubled.

"Next you two will be studying geology," he said in a flat tone that almost sounded robotic. "Which is the study of the earth's physical features. You will learn about things such as mountains, islands, and volcanoes. You will learn about different kinds of rocks and how those rocks can be changed by weathering. You will learn about how the surface of the earth shifts . . . and moves . . . and how the . . . and the . . . geo . . . blubby . . . gossterf . . ."

Blaine and Tracey looked at each other with concerned glances. Uncle Cecil had slowed down so much he wasn't even using real words anymore. He must really be worried about Summer.

"She is going to be okay, Uncle Cecil," Tracey said with kind optimism. "We will figure out what happened to Summer, and we will figure out how to help her."

"Zilla . . . mackry . . . whooda . . . ohhh, wait a second!" Cecil suddenly went from mumbling mode to clairvoyance mode. "That has to be it! Right? Don't you two agree?"

"Agree with what?" the twins thought. Their uncle hadn't actually shared a clear thought.

"It's similar to what happened to Blaine earlier this week!"

"What are you talking about, Uncle Cecil?" Blaine asked. "What happened to me earlier this week?"

"Persactly!" Cecil responded. "That is why you were acting foggy and aloof."

"Huh?" Both twins responded.

"But that doesn't explain how she ended up here on Pecan Street," Cecil continued, pretty much just having a conversation with himself. "Her location, however, is really just a secondary matter. She could have very well just absent-mindedly used the harness and three-ringed carabiner that I gave her to get here. Though that would be somewhat miraculous because she would have to remember those coordinates. Unless, of course, she recorded the coordinates or wrote them down somewhere, which is highly possible considering how organized she is. Point in fact, she has been like that since junior high, which is something I always liked about her but have yet to attain for myself. If she did zip here, then maybe Ulysses—"

"Uncle Cecil!" the twins interjected. "What are you talking about? What happened to Summer?"

"Her memory has been erased," the redheaded scientist informed with sorrow on his face.

"Her memory has been erased?" Tracey asked. "How can you be so sure?"

"I'm not sure," Cecil responded. "But at this point it seems to be the most plausible explanation."

He looked at his nephew. "Blaine, a few days ago, I found you absentmindedly wandering the sidewalks, much in the same fashion that we found Summer today."

"You did?" Blaine responded. "I don't remember that. Why would I have been wandering the sidewalks?"

"I don't know," Cecil answered. "You disappeared while we

were being chased by Old Man Grusher's dog. I thought you had zipped off to your first earth science location without Tracey, but you hadn't. You were still here. When I found you, however, you were acting very forgetful, and you couldn't clearly explain where you had been. It was like someone had meddled with the memory in your brain. And now, today, we find Summer in the same state, except her condition seems much worse than yours did."

Cecil reached up and put his thumbs on Blaine's temples and looked into his eyes like a doctor might. "Are you feeling okay now? Are you remembering everything you need to remember?"

"I think so," Blaine responded.

"Yes, you seem fine," Cecil agreed, dropping his hands. "You normalized rather quickly, but I don't think this is going to be the case with Summer. I think her memory is completely wiped clean."

"But how could that happen?" Tracey asked. "Who would be evil enough to take someone and wipe their memory away?"

"Only one person comes to mind," Cecil said.

Blaine and Tracey both immediately knew about whom their uncle was referring. "The Man with No Eyebrows!" they exclaimed together.

The scientist nodded.

The three Sassafrases stood silently in the basement for a while, thinking about all the swirling possibilities and implications. Suddenly, Cecil's face lit up. He looked toward his lab assistant.

"Linc-Dawg! Do you remember that memory project we were working on a few months ago? You know, the one where I was going to transfer all the contents of my brain into a canister so I could literally share my thoughts with others?"

The prairie dog bobbed his head in a sort of nod.

"Summer was working on that project with us, wasn't she?

I think she was, right?" Cecil was still excited, but also confused because he wasn't sure if he was remembering right.

"Linc-dawg, I know you didn't love that project, but Summer did. She thought it was a great idea. So much, in fact, that I'm pretty sure she was willing to try it out. Right? Isn't that what happened?"

Now the twins couldn't tell if President Lincoln was still nodding or if he was shrugging his shoulders.

"If that is the case," Cecil continued, with a hopeful shout, "somewhere in this basement, we very well might have a canister with the contents of Summer's brain in it! If we do have that canister and can find it, we can give Summer back her memory. And if we can give Summer back her memory, then she will be herself again!"

Blaine and Tracey found themselves just as hopeful as their uncle, but as they looked around at the messy piles down in the basement, some of that hope began to fade away. Did this canister even exist?

And if it did exist, how were they going to be able to find anything in all of this junk?

# Stay in Touch with the Sassafras Twins!

**The adventure doesn't have to end just because you've finished the book!** Connect with the twins and the other characters of the series through the Sassafras Science blog. You'll find articles in which:

- ☆ Uncle Cecil explains how to make a rocket at home;
- ☆ The Prez shares his wrap-up videos;
- ☆ Atif explains what causes irridescent clouds;
- ☆ Hawk Talons shows how to make it snow indoors.

Plus, Blaine and Tracey regularly pop in to say hi and share their thoughts. The Sassafras Science blog is *the place* to get to know the characters of the series!

The Sassafras twins would also love to keep in touch with you through their Facebook page. They share updates about future books, fun science-related activities, and cool nature news!

# Visit SassafrasScience.com and click on the "Blog" tab to discover more!